MW00843437

LABORATORY MANUAL TO ACCOMPANY

ELECTRONIC COMMUNICATIONS
A Systems Approach

by JEFFREY S. BEASLEY, JONATHAN D. HYMER, and GARY M. MILLER

With contributions by

JEFFREY S. BEASLEY

JONATHAN D. HYMER

MARTIN S. MASON

KENNETH J. MILLER

MARK E. OLIVER

DAVID H. SHORES

PEARSON

Boston   Columbus   Indianapolis   New York   San Francisco   Upper Saddle River
Amsterdam   Cape Town   Dubai   London   Madrid   Milan   Munich   Paris   Montreal   Toronto
Delhi   Mexico City   Sao Paulo   Sydney   Hong Kong   Seoul   Singapore   Taipei   Tokyo

Editorial Director: Vernon R. Anthony
Acquisitions Editor: Lindsey Prudhomme Gill
Editorial Assistant: Amanda Cerreto
Director of Marketing: David Gesell
Senior Marketing Coordinator: Alicia Wozniak
Senior Managing Editor: JoEllen Gohr
Senior Project Manager: Rex Davidson
Senior Operations Supervisor: Pat Tonneman
Creative Director: Andrea Nix
Art Director: Jayne Conte
Cover Designer: Suzanne Behnke
Cover Image: Fotolia
Full-Service Project Management: Peggy Kellar/iEnergizer Aptara®, Inc.
Composition: Aptara® Inc.
Printer/Binder: Edwards Brothers Malloy
Cover Printer: Edwards Brothers Malloy

Copyright © 2014 by Pearson Education, Inc. All rights reserved. Manufactured in the United States of America. This publication is protected by Copyright, and permission should be obtained from the publisher prior to any prohibited reproduction, storage in a retrieval system, or transmission in any form or by any means, electronic, mechanical, photocopying, recording, or likewise. To obtain permission(s) to use material from this work, please submit a written request to Pearson Education, Inc., Permissions Department, One Lake Street, Upper Saddle River, New Jersey 07458, or you may fax your request to 201-236-3290.

Many of the designations by manufacturers and sellers to distinguish their products are claimed as trademarks. Where those designations appear in this book, and the publisher was aware of a trademark claim, the designations have been printed in initial caps or all caps.

10 9 8 7 6 5 4 3 2 1

ISBN 10:   0-13-301066-X
ISBN 13: 978-0-13-301066-4

PREFACE

This laboratory manual is intended for use with *Electronic Communications: A Systems Approach* by Jeff Beasley, Jonathan Hymer, and Gary Miller. We have attempted to create a lab manual that significantly expands and adds depth to the presentation contained within the accompanying text. To this end, we have devoted considerable effort to crafting introductions for each lab activity that enhance the discussion of concepts first introduced in the companion work. We encourage students and instructors alike to read text and lab manual side-by-side and in advance of each laboratory session to enhance their understanding of fundamental communications concepts.

The thirty-four experiments contained herein constitute a mix of hardware- and software-based, circuit- and systems-level treatments. Taken together, these experiments provide complete coverage of the most important ideas embodied in the field of electronic communication. This manual is significantly updated from the laboratory manual for the 9th edition of *Modern Electronic Communication*, the book from which the systems approach text was derived. In particular, the digital communications concepts illustrated in Experiments 14 through 18 make use of the Arduino microcontroller to present a unique approach to communication studies at the systems level. The Arduino programming code for these experiments, as well as pictorial diagrams for interfacing the microcontroller to components on an external prototype board, is provided in the Appendix. Also new to this manual are six experiments focused on concepts crucial to computer networks. It goes without saying that computer-based communication has revolutionized commerce and information technology alike. We are excited to include hands-on, lab-based coverage of this crucial area of electronics that significantly enhances the text presentation.

At no time in history has computer processing been as powerful and as affordable as it is now. This power is brought to bear in solving one of the most intractable problems facing instructors in communications-related topics: the cost of relevant laboratory equipment. In particular, the spectrum analyzer is an indispensable tool in communications work. This piece of test equipment has historically been prohibitively expensive for most educational institutions, but its importance to communications-related occupations cannot be overstated. Fortunately, high-quality spectrum analyzers that make extensive use of digital signal-processing (DSP) techniques are now available at a fraction of the cost of their predecessor units, and now there is practically no excuse for not having modern spectrum analyzers available for student use in the laboratory. In particular, we recommend the Rigol Technologies model DSA-815-TG instrument, which features a frequency range of 9 kHz to 1.5 GHz with 100-Hz resolution bandwidth. This product is accurate, full-featured, and affordable, and its user interface is very similar to that of analyzers from other manufacturers, thus making it an ideal platform for training students on the instruments they are likely to encounter on the job. So powerful is the Rigol DSA-815 that five experiments in this manual have been written around it specifically. The analyzer equipped with tracking generator option sells for less than $1500, about one-tenth the cost of previous-generation instruments from other manufacturers. At this price point, it is possible to equip an entire laboratory with high-quality test equipment for student use rather than to rely solely on a single analyzer for demonstration purposes. More information about this instrument can be found at the manufacturer's website at http://www.rigolna.com. We are well aware of the budget constraints faced by educational institutions, but the availability of professional-grade equipment at the price of the Rigol DSA-815 is a game-changer in communications education. We advocate moving mountains and otherwise shaking the procurement tree as hard as possible to make these analyzers available for students—their employment prospects and competence on the job can only be enhanced if they have the opportunity to avail themselves in the laboratory of high-caliber test equipment such as this.

Other equipment required to complete labs is typical of that found in student laboratories, and instructors are encouraged to modify procedures as appropriate for their situations. Experiments 29 and 30 call for the use of various filters and antennas. We recommend the use of filters from Mini-Circuits (http://www.mini-circuits.com) and the LPY915 printed-circuit board antenna from Ramsey Electronics (http://www. ramseykits.com). Experiments 32 and 33 require some rather expensive hand tools and supplies, but the fiber-optic cabling and connectorization skills introduced are in high demand by industry and are definitely worth covering if time and finances permit. Many of the tools and supplies are available from Fiber Instrument Sales, Inc. (http://www.fiberinstrumentsales.com). Part numbers for most of the equipment and supplies have also been included.

Although some experiments may initially appear quite long, our experience confirms that each experiment can be completed within a single laboratory period, provided that students are adequately prepared. Preparation is an essential job skill, and this skill can be taught in part by encouraging students to read through lab assignments in advance of each class meeting. In some cases, having circuits or other materials fabricated ahead of time will eliminate the need for extensive setup and construction during the laboratory period and will minimize the potential for the kinds of problems attendant to radio-frequency (RF) circuits. If resources such as departmental technical staff are not available, then exercises should be extended to multiple lab periods to permit student assembly. The Instructor's Manual that accompanies the text provides further details regarding the advance preparation needed to perform some of the labs in this manual. Still other labs, or certain procedure steps within them, may be better suited for completion as part of an instructor-led demonstration rather than by students, particularly if lab time is at a premium.

In our collective experience of teaching electronics and other engineering-related topics, we have found that the discipline truly comes alive for students when they engage with the subject matter personally in the laboratory context. The joy of electronics surely rests with its hands-on nature and with the opportunities it presents for tinkering and experimentation. We too have taken great joy in crafting experiments for this manual that, we believe, will make the field come alive in a way that no book or classroom lecture alone possibly can. With the experiments represented herein we hope to have demonstrated that at no time in history has the communications electronics field presented as much opportunity for discovery and innovation as it presents right now.

ACKNOWLEDGMENTS

We would first like to thank Rigol Technologies and sales manager Steve Barfield for making a DSA-815-TG spectrum analyzer available for the development of several new and unique laboratory activities. Again, the availability of this instrument opens up many opportunities for students, and we are excited to be able to integrate its use in so many ways. Thanks also to Mark Oliver, Jeff Beasley, and David Shores for their contributions to the 9th edition of *Laboratory Manual to Accompany Modern Electronic Communication*, some of whose content we have repurposed for this work.

Special thanks are also due to Leslie Lahr of Aptara for her insightful developmental editing work and for her tireless attention to the details of manuscript preparation. Together with project managers Peggy Kellar, also of Aptara, and Rex Davidson of Pearson, they have made a daunting project doable. Thanks also to senior acquisitions editor Lindsey Prudhomme Gill and editorial director Vernon Anthony of Pearson for making the resources available to produce a text and lab manual that, together, address the realities faced by those entering the communications field in the second decade of the 21st century. Finally, thanks to all our students, past and future, who remind us that it is the joy of discovery that makes teaching as much a calling as it is a vocation.

Jonathan D. Hymer
Martin S. Mason
Kenneth J. Miller
Mt. San Antonio College
Walnut, California

CONTENTS

Experiment 1: DECIBEL MEASUREMENTS IN COMMUNICATIONS 1

Experiment 2: WAVEFORMS IN THE TIME AND FREQUENCY DOMAINS 7

Experiment 3: INTRODUCTION TO SPECTRUM ANALYSIS 17

Experiment 4: UPCONVERSION AND DOWNCONVERSION 27

Experiment 5: FREQUENCY MODULATION: SPECTRUM ANALYSIS 31

Experiment 6: RADIO-FREQUENCY AMPLIFIERS AND FREQUENCY MULTIPLIERS 45

Experiment 7: COLPITTS RF OSCILLATOR DESIGN 53

Experiment 8: HARTLEY RF OSCILLATOR DESIGN 59

Experiment 9: PRINCIPLES OF NONLINEAR MIXING 65

Experiment 10: SIDEBAND MODULATION AND DETECTION 71

Experiment 11: FM DETECTION AND FREQUENCY SYNTHESIS USING PLLs 79

Experiment 12: GENERATING FM FROM A VCO 85

Experiment 13: PULSE AMPLITUDE MODULATION 89

Experiment 14: TIME-DIVISION MULTIPLEXING 95

Experiment 15: PULSE-WIDTH MODULATION AND DETECTION 101

Experiment 16: INTRODUCTION TO ANALOG-TO-DIGITAL CONVERSION (ADC) AND DIGITAL-TO-ANALOG CONVERSION (DAC) 107

Experiment 17: PULSE CODE MODULATION AND SERIAL DATA PROTOCOLS 115

Experiment 18: FREQUENCY SHIFT KEYING MODULATION AND DEMODULATION 125

Experiment 19: MODEM COMMUNICATIONS 133

Experiment 20: ROUTER CONFIGURATION 137

Experiment 21: WIRELESS COMPUTER NETWORKS 143

Experiment 22: PLANNING AND DESIGNING LOCAL-AREA NETWORKS (LANs) 149

Experiment 23: LOCAL-AREA NETWORK (LAN) TROUBLESHOOTING 157

Experiment 24: BINARY AND IP ADDRESSING 163

Experiment 25: STANDING-WAVE MEASUREMENTS OF A DELAY LINE 171

Experiment 26: USING CAPACITORS FOR IMPEDANCE MATCHING 181

Experiment 27: SMITH CHART MEASUREMENTS USING THE MULTISIM NETWORK ANALYZER 185

Experiment 28: MULTISIM—IMPEDANCE MATCHING 193

Experiment 29: **SCALAR NETWORK ANALYSIS AND VOLTAGE STANDING-WAVE RATIO (VSWR) MEASUREMENTS 201**

Experiment 30: **ANTENNA POLAR PLOTS AND GAIN CALCULATIONS 213**

Experiment 31: **FIBER OPTICS 223**

Experiment 32: **FIBER-OPTIC CABLE SPLICING 227**

Experiment 33: **FIBER-OPTIC CABLE CONNECTORIZATION 235**

Experiment 34: **FIBER-OPTICS COMMUNICATION LINK 247**

Appendix **ARDUINO PROGRAMMING CODE AND BREADBOARD LAYOUTS 253**

DECIBEL MEASUREMENTS IN COMMUNICATIONS

OBJECTIVES:

1. To become acquainted with the use of National Instruments Multisim in simulating dB (decibel) measurements in communications circuits.
2. To understand dB measurement using the multimeter set to the dB mode.
3. To explore the characteristics of a T-type passive attenuator circuit.
4. To explore the procedure for setting the dB levels in a system.

REFERENCE:

Refer to Section 1-2 in the text.

EQUIPMENT:

None

COMPONENTS:

None

INTRODUCTION:

This experiment introduces the techniques for making audio signal level measurements using Multisim simulations. Obtaining signal level measurements and measuring signal path performance are common maintenance, installation, and troubleshooting practices in all areas of communications. Communication networks require the maintenance of proper signal levels to ensure minimum line distortion and crosstalk. This section examines the techniques for making dB (decibel) measurements on various passive circuits and for setting system levels.

PROCEDURE:

Part I: Measuring the Insertion Loss Provided by Passive Resistive Attenuator Circuits

You will make dB level measurements on passive attenuator circuits and will perform a system-wide calibration of the levels.

1. The Multisim implementation of a test circuit for making dB measurements is provided in Figure 1-1. This circuit contains an ac signal source, a 600 Ω load, and two multimeters. The top multimeter (XMM1) is measuring the dB level and the bottom multimeter (XMM2) is being used to measure the voltage across the load. Construct the circuit and set the ac voltage source to 2.188 V at 1 kHz. Measure the voltage and dB levels across R2. Record your value. Note: Make sure you set the multimeter to measure ac.

_____ V

_____ dB

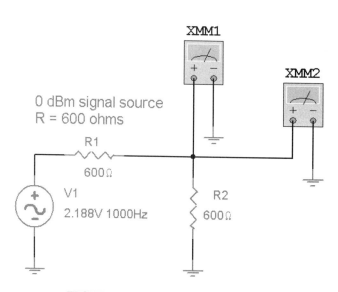

FIGURE 1-1   Multisim test circuit.

2. Construct the circuit shown in Figure 1-2.

This circuit includes a 600 Ω, 0-dBm voltage source and a 600-Ω T-type attenuator that has been inserted into the path between the signal source and the 600-Ω load resistance. Measure the amount of insertion loss provided by the T-type attenuator. This requires that you measure the dB levels at both the input and output of the attenuator. The insertion loss will be the difference in the two measurements.

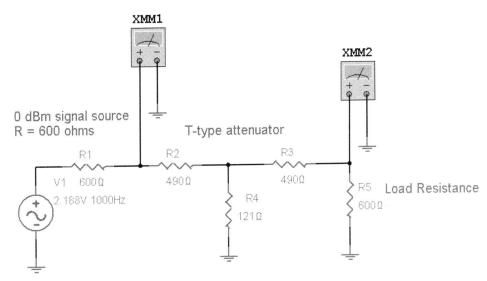

FIGURE 1-2 T-type attenuator circuit with 600-Ω source and equivalent load resistance.

dB level measured at the attenuator input _____

dB level measured at the attenuator output _____

insertion loss (output − input) _____

3. Repeat step 2 for the resistor values provided for the T-type attenuator values listed in Table 1-1.

TABLE 1-1   Input and Output Levels and Insertion Loss for a T-type Attenuator.

R2	*R3*	*R4*	INPUT LEVEL (**dB**)	OUTPUT LEVEL (**dB**)	INSERTION LOSS (**dB**)
230	230	685			
69	69	258			
588	588	120			
312	312	422			
563	563	38			

4. Remove the signal source and connect multimeter XMM 1 to the input of the T-type attenuator at R2, as shown in Figure 1-3. Measure the equivalent resistance of the attenuator circuit and verify that the measured resistance for the attenuator values listed in Table 1-1 are each equal to 600 Ω.

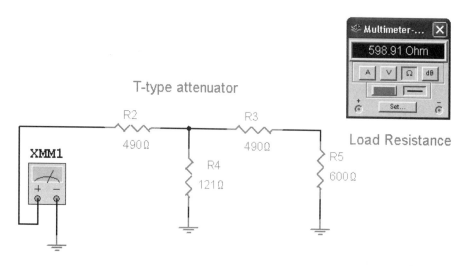

FIGURE 1-3 Circuit for determining equivalent resistance of T-type attenuator.

Repeat step 4 for the resistor values provided in Table 1-2. Make sure in each case that the T-type attenuator is terminated with a 600-Ω resistor. The resistor $R5$ is the termination resistor in both Figures 1-2 and 1-3.

TABLE 1-2   Input Resistance for T-type Attenuator for Resistor Combinations of Table 1-1.

$R2$	$R3$	$R4$	INPUT RESISTANCE (Ω)
230	230	685	
69	69	258	
588	588	12	
312	312	422	
563	563	38	

Part II: Setting System-Wide dB Levels

This exercise demonstrates the procedure for setting dB levels in a communication system. This example is for a broadcast facility where the audio levels are being set to a predefined level throughout the system. The system is shown in Figure 1-4. The audio outputs for a tone generator, production audio, VCR1, and a satellite feed are shown. These devices are inputted into a passive rotary switch. The output of the switch feeds the station studio transmitter link (STL). The level's input to the rotary switch (TP1) must be set to 0 dBm. The level to the STL must be set to +8 dBm (TP2). A 1-kHz tone is used to calibrate the system. Level adjustment is provided by the virtual potentiometer connected in the feedback paths for the operational amplifiers.

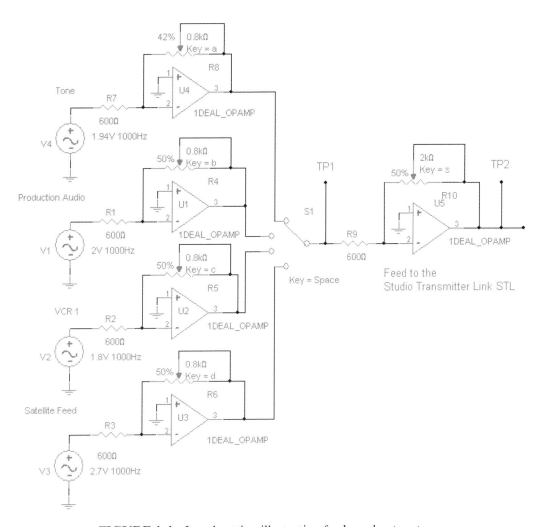

FIGURE 1-4 Level-setting illustration for broadcast system.

5. Set the levels according to Table 1-3.
6. Once you have completed the settings, demonstrate to your lab instructor that all levels have been properly set. Make sure that the multimeter is properly terminated with 600 Ω when making your measurements.

TABLE 1-3   Signal Levels for Sources in Broadcast System.

DEVICE	OUTPUT LEVEL	LEVEL CONTROL KEY
Tone	0 dBm	**A** increase, **a** decrease
Production Audio	0 dBm	**B** increase, **b** decrease
VCR1	0 dBm	**C** increase, **c** decrease
Satellite Feed	0 dBm	**D** increase, **d** decrease
Rotary Switch	0 dBm	None
STL	+8 dBm	**S** increase, **s** decrease

QUESTIONS:

1. Explain how the voltage measured across R2 in Figure 1-1 equates to 0 dBm. (Hint: Recall that the output impedance of the generator is 600 Ω, that R1 and R2 form a series circuit, and that power levels must be expressed as rms values.)

2. Imagine you are in a job interview. Use your answer to question 1 to explain concisely the effect of generator impedance on system levels. Why it is important to be aware of the output impedance of the generator? Is it possible for the generator to have an output impedance other than 600 Ω? If so, would use of the dBm terminology be appropriate?

3. What is the purpose of an attenuator?

4. Referring to Section 1-2 in the text as necessary, explain the following:

   (a) Why the decibel formula for voltages must be modified when the input and output impedances are different.

   (b) Why there is no change to the dB power formula for differing input and output impedances.

5. If an amplifier has 6 dB of gain, how much is the power increased? _____. How much is the voltage increased, assuming equal input and output impedances? _____.

6. If you measured –47 dBm at the input of an amplifier and –24 dBm at the output, what would the gain in dB be? _____ dB. Why is this gain not in dBm?

7. In an amplifier that has a power gain of 6 dB, what is the voltage gain in dB? Why?

8. In Part II, did the position of the rotary switch make any difference when setting the audio levels? Explain.

9. Do you need to add a 600-Ω termination resistor when measuring the audio level from the output of the rotary switch? Explain.

10. Do you need to add a 600-Ω termination resistor when measuring the audio level feeding the STL? Explain.

WAVEFORMS IN THE TIME AND FREQUENCY DOMAINS

OBJECTIVES:

1. To become acquainted with the analysis of waveforms in the time and frequency domains.
2. To become familiar with the use of National Instruments Multisim software and the operation of its spectrum analyzer and Tektronix oscilloscope simulations.
3. To measure and analyze the spectral content of a sinusoid, a triangle wave, and a square wave.

REFERENCE:

Refer to Section 1-3 in the text.

EQUIPMENT:

None. This assignment uses Multisim.

COMPONENTS:

None.

INTRODUCTION:

The purpose of this experiment is to introduce the spectrum analyzer and to demonstrate its use in identifying the frequency components of several types of signals. In this experiment we will use a software simulation. In a later experiment you will have the opportunity to work with a modern, hardware-based analyzer capable of operation into the gigahertz region.

Electronic signals can be viewed in either the *time domain* or the *frequency domain*. An oscilloscope visually represents signals in the time domain by displaying amplitude variations with respect to time. The frequency of one cycle of a sine wave is the reciprocal of its period and is easy to determine with an oscilloscope because it is the only frequency present. However, more complex waveforms such as square or triangle waves, or waves having an impulse-like nature, can be shown to consist of multiple sine and/or cosine waves, each with a definite frequency and amplitude relationship with respect to the others. The oscilloscope is not well suited to displaying the frequency characteristics of such complex signals because the harmonics are effectively hidden from view.

A frequency-domain representation shows signal amplitudes as a function of frequency rather than time. A *spectrum analyzer* is a test instrument that provides a frequency-domain display of all the components of signals applied to its input. The spectrum analyzer is an invaluable tool in communications work because it reveals not only the amplitudes and frequency relationships of desired signals but also the presence and strengths of undesired and potentially interfering signals. Experiment 3 will introduce the operation of a hardware spectrum analyzer designed for radio frequencies. This experiment uses a simulation in Multisim to illustrate the principles of spectrum-analyzer operation.

Because signals encountered in electronic communications systems are most often repetitive and complex (i.e., consisting of multiple frequency components with distinct phase relationships), communications engineers make extensive use of a mathematical technique known as *Fourier analysis* to resolve the fundamental and harmonic frequency components of complex signals. Fourier analysis readily allows for conversion from the time to the frequency domain and vice versa. The Fourier series of a repetitive waveform represents periodic functions of voltage or current as an infinite series of sine or cosine functions. In its general form, the Fourier series is expressed as

$$f(t) = A_0 + \sum_{n=1}^{\infty} A_n \cos(n\omega_n t + \phi_n) + \sum_{n=1}^{\infty} B_n \sin(n\omega_n t + \phi_n).$$

Fortunately, most waveforms can be represented in much easier terms. The easiest waveform to represent mathematically is the sine wave itself.

$$e(t) = E_0 + E_{max} \cos(\omega t + \phi),$$

where E_o = dc offset
E_{max} = peak value of the sine wave
ω = frequency, radians per second
ϕ = initial phase angle,

Other popular waveforms that can be written as a Fourier series are the square wave and triangle wave shown, respectively, in Figure 2-1(a) and (b).

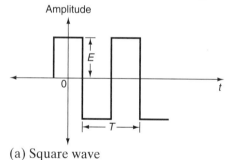

$$e(t) = \frac{4E}{\pi}\left[\cos(\omega_1 t + \phi) + \frac{1}{3}\cos(3\omega_1 t + \phi) + \cdots + \frac{1}{n}\cos(n\omega_1 t + \phi) \right]$$

where $T = 1/f = 2\pi/\omega_1$ and n = odd integers only.

(a) Square wave

$$e(t) = \frac{8E}{\pi^2}\left[\cos(\omega_1 t + \phi) + \frac{1}{9}\cos(3\omega_1 t + \phi) + \cdots + \frac{1}{n^2}\cos(n\omega_1 t + \phi) \right]$$

where $T = 1/f = 2\pi/\omega_1$ and n = odd integers only.

(b) Triangle wave

FIGURE 2-1   Popular waveforms and their Fourier series.

If the Fourier series of a waveform is known, its spectral content can be represented on a *spectrum diagram*. A spectrum diagram is a sketch of voltage (or current) versus frequency in which its individual harmonic amplitudes are plotted as vertical lines (arrows) at each harmonic frequency. For example, for a 60-Hz sawtooth waveform (see Figure 2-2):

$$e(t) = \frac{2E}{\pi}\left[\sin(377t) - \frac{\sin(2 \times 377t)}{2} + \frac{\sin(2 \times 377t)}{3} + \cdots + \frac{(-1)^{n+1}\sin(n \times 377t)}{n}\right].$$

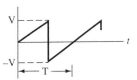

FIGURE 2-2 Sawtooth waveform.

Figure 2-3 shows the spectrum of this signal if $E = 10$ V.

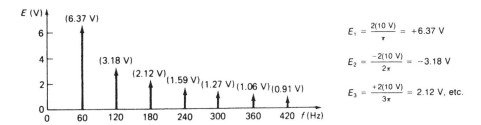

FIGURE 2-3 Spectrum diagram for sawtooth waveform with initial voltage of 10 V.

Notice that if the peak voltage is negative, it is still displayed as positive, since the negative sign merely denotes a 180° phase shift of that sinusoidal term. A spectrum diagram can be displayed for an unknown signal using a spectrum analyzer. Often, the spectrum analyzer's display is more useful in communication applications than is the more popular oscilloscope display of waveforms.

Software simulations, as well as modern hardware spectrum analyzers, make extensive use of the *fast Fourier transform*, an approximation of the full Fourier series that lends itself to implementation in software. Many math software packages have an FFT function, as do modern digital sampling oscilloscopes. An oscilloscope with "FFT math" software built into it is effectively functioning as a spectrum analyzer when it performs the FFT operation.

PROCEDURE:

Part I: Analyzing a Sine Wave with the Multisim Spectrum Analyzer

1. Assemble the test setup shown in Figure 2-4. The setup shows an oscilloscope (time-domain instrument) and a spectrum analyzer (frequency-domain instrument), each connected to the output of a function generator. The function generator will be used to generate the waveforms being analyzed.

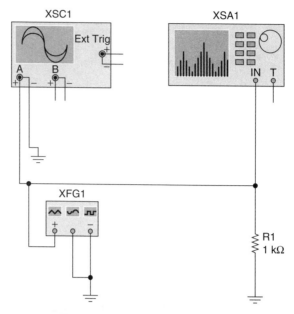

FIGURE 2-4 Multisim test setup.

If you are unfamiliar with Multisim, please ask the instructor or a classmate for assistance. Note in particular that test instruments are on the vertical menu located to the right of the workspace and that individual components such as the 1-kΩ resistor shown in the figure are placed from the menu located above the work area. To run the simulation, press the toggle switch located at the top right of the Multisim window as shown in Figure 2-5.

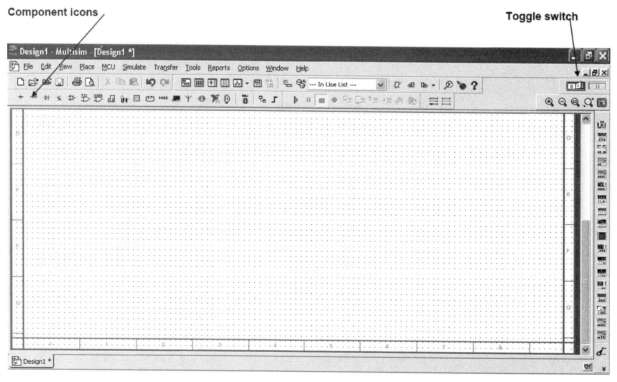

FIGURE 2-5 Multisim window showing locations of components and test equipment icons.

2. Double-click on the function generator. You should see the panel display of the function generator shown in Figure 2-6.

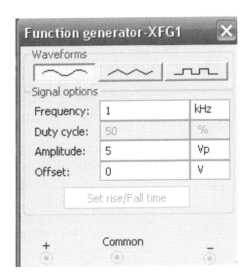

FIGURE 2-6 Function generator front panel.

3. The function generator is capable of producing sine, triangle, or square waves. Click on the sine wave. Set the frequency to 1 kHz, the amplitude to 5 V peak, and the offset to 0 V.

4. Start the simulation by pressing the toggle switch located at the upper right-hand side of the window. Double-click on the oscilloscope icon and verify that you are outputting a 1-kHz sine wave. Next, double-click on the spectrum analyzer. You should see a display similar to the one shown in Figure 2-7. Change the settings to match those shown in the figures using the guidelines shown in the following substeps.

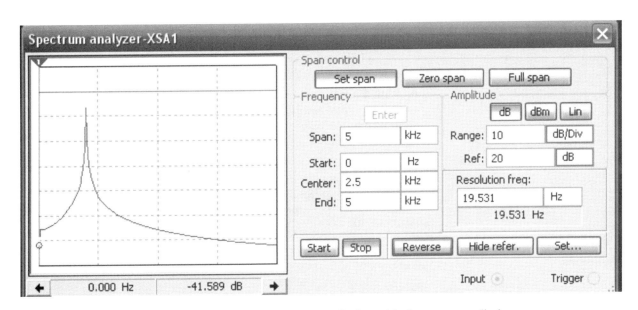

FIGURE 2-7 Spectrum analyzer display with sine wave applied.

a. Click on the Set span button. This provides the user with the ability to set the Span and Frequency controls. Set the span to 5 kHz, the start frequency to 0 Hz, and the end frequency to 5 kHz, then click on "Enter." The center frequency is automatically calculated. The span-setting controls the frequency range displayed by the spectrum analyzer. (Note that this step is unique to the Multisim spectrum analyzer. On most hardware analyzers, the center frequency is entered directly, and the frequency span and/or start and stop frequencies can be directly entered also.)

b. Set the amplitude to dB and the range to 10 dB/div. Setting the amplitude to dB produces outputs that are in relative terms, and the range determines the decibel increment by which each major division is reduced from the top, or reference level.

c. Set the Resolution frequency to 19.531 Hz. This setting defines how many data points are to be used by the spectrum analyzer to display the frequency spectra. The smaller the resolution frequency, the better and more accurate the display. The smallest resolution frequency for the simulated spectrum analyzer is determined from the following equation:

$$\text{Minimum resolution frequency} = \frac{f\_end}{1024},$$

where f_end is the end frequency of the span. The number 1024 comes about because the simulation uses 10 bit binary resolution, thus $2^{10} = 1024$. Because in step 4(a) the end frequency was set to 5 kHz, the minimum frequency resolution is 4.8828 Hz:

$$\text{Minimum resolution frequency} = \frac{f\_end}{1024} = \frac{5000}{1024} = 4.8828 \text{ Hz.}$$

Also, for the simulation to work properly, the resolution frequency specified must be an even multiple of the minimum resolution frequency. Thus, the 19.531Hz resolution frequency specified is four times the calculated minimum of 4.8828 Hz. Also note that the minimum resolution frequency does not generally need to be calculated with a hardware spectrum analyzer. The operating parameters would automatically be determined from the operating settings.

d. With a calculator, use the relationship given in step 4(c) to determine the minimum resolution frequency for the following end frequencies and enter the results in Table 2-1:

TABLE 2-1 Calculated Minimum Resolution Frequencies for Selected End Frequencies.

MINIMUM END FREQUENCY	MINIMUM RESOLUTION FREQUENCY
100 kHz	
455 kHz	
108 MHz	
433 MHz	

Part II: Spectral Analysis of a Square Wave

5. Change the output of the function generator to a 1-kHz square wave with a 50% duty cycle (remember, the definition of a square wave is that it has a 50% duty cycle), a 10-V amplitude, and 0-V offset voltage. Double-click on the spectrum analyzer and change the span to 15 kHz. Start the simulation and observe the results on the spectrum analyzer.

6. Identify the frequency of the harmonic components being displayed, their dB levels, and the harmonic by number. Do this by placing the slider directly over each harmonic. You should be seeing the first through thirteenth harmonics. The first harmonic (also known as the fundamental) is 1 kHz in this example, the third harmonic (which is the first odd harmonic) is 3 kHz, and so on. Recall from Section 1-3 in the text that square waves contain content from the odd harmonics only; therefore, there should be little or no energy represented in the columns in Table 2-2 associated with the even harmonics. Record the results in Table 2-2.

TABLE 2-2 Amplitudes of Fundamental and First Seven Odd Harmonics of Square Wave.

HARMONIC	1ST	2ND	3RD	4TH	5TH	6TH	7TH	8TH	9TH	10TH	11TH	12TH	13TH
Frequency													
dB value													
Harmonic													

Part III: Spectral Analysis of a Triangle Wave

7. Change the output of the function generator to a 1-kHz triangle wave with a 50% duty cycle, a 10-V amplitude, and 0-V offset voltage. Double-click on the spectrum analyzer and change the span to 15 kHz. Start the simulation and observe the results on the spectrum analyzer.

8. Identify the frequency of the components (harmonics) being displayed, their dB levels, and harmonic number. Do this by placing the slider over each harmonic. You should be seeing the first through thirteenth harmonics. As before, the first harmonic is 1 kHz, the third harmonic is 3 kHz and so on. Record the results in Table 2-3.

TABLE 2-3 Amplitudes of Fundamental and Harmonics of Triangle Wave.

HARMONIC	1ST	2ND	3RD	4TH	5TH	6TH	7TH	8TH	9TH	10TH	11TH	12TH	13TH
Frequency													
dB value													
Harmonic													

Part IV: FFT Math Simulation

Among the available test instruments in Multisim is a simulated Tektronix model TDS 2024 digital storage oscilloscope. Currently, this model is a widely used, industry-standard bench instrument. As described in the introduction, both the hardware instrument and the simulation have an "FFT math" feature, which causes the oscilloscope effectively to function as a spectrum analyzer.

9. Place the simulated Tektronix oscilloscope on your workspace and connect channel 1 to the function generator. (The instrument is represented by the bottom instrument icon in the vertical column on the right-hand side of your work space as shown in Figure 2-8.) You may delete the generic oscilloscope from your test setup or connect the Tektronix oscilloscope in parallel with it.

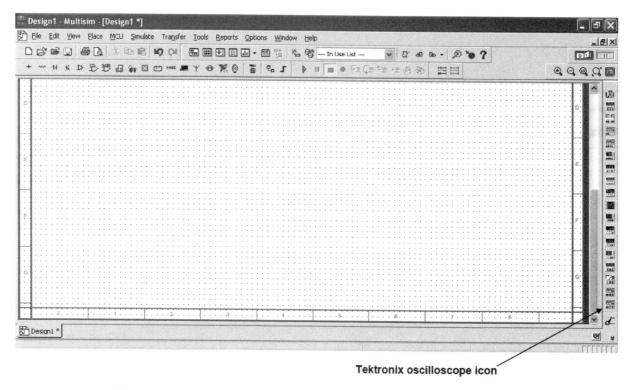

Tektronix oscilloscope icon

FIGURE 2-8   Location of instrument icon for Tektronix oscilloscope in Multisim workspace.

10. Double-click on the function generator and, if necessary, return its settings to those shown in step 2 and Figure 2-6.

11. Double-click on the Tektronix oscilloscope. You should see an image similar to that of Figure 2-9. Note, however, that the simulation is so exact that you have to remember to depress the virtual power switch, located at the left-hand side of the instrument, before you will see anything on the screen.

12. Start the simulation. Adjust the CH 1 volts/div knob, vertical position controls, and sec/div knob as needed to view several cycles of the sine wave. Note in particular how exact the simulation is: This simulation comes as close to operation of the actual instrument as possible short of having the device in front of you. Take some time to experiment with the other controls to get a feel for how an oscilloscope of this type is used.

13. Press the pink "math menu" button located between the yellow CH 1 and blue CH 2 menu buttons. The five buttons immediately to the right of the screen are known as "soft keys" because their functions change depending

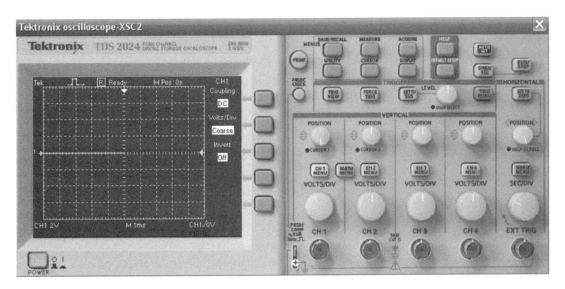

FIGURE 2-9 Simulated Tektronix TDS 2024 oscilloscope.

on the instrument state. Press the uppermost soft key (the word "operation" should appear on the screen to the left of it) several times until "FFT" is displayed. In the space below describe what you see.

14. Change the function generator settings to produce a 1-kHz square wave. In the space below describe what you see.

15. As in step 12, take some time to experiment with the functionality of the simulated oscilloscope. Note in particular that the vertical scale (now shown in dB/div at the lower left) can be adjusted with the volts/div knob, and the center frequency can be adjusted with the horizontal position and time base controls.

QUESTIONS

1. Refer to step 4 and Table 2-1 for this question. How many frequency components does the spectrum analyzer display for the sine wave? Use the slider located directly below the display to determine the frequency of the displayed signal. Is this what you expected for a sine wave? Explain in the space below.

2. Refer to your data in Table 2-2 to answer the following questions:

   (a) Which harmonic frequency has the greatest amplitude (dB) value?

   (b) When a signal is reduced in amplitude from that of a reference, it is said to be "down" by the number of decibels (dB) indicated on the display. How many "dB down" is the third harmonic relative to the first harmonic?

   (c) How many dB down is the seventh harmonic relative to the first harmonic?

3. Refer to your data in Table 2-3 to answer the following questions:

   (a) Which harmonic frequency has the greatest amplitude (dB) value?

   (b) How many dB down is the fifth harmonic relative to the first harmonic?

   (c) How many dB down is the eleventh harmonic relative to the first harmonic?

4. Refer to your work in step 13. Explain how the display is or is not consistent with what you would expect to see for a sine wave.

5. Refer to your work in step 14. Did the spikes on the screen reflect what you would expect to see for the frequency spectrum for a square wave? Explain your answer; include an explanation of how you can determine the frequencies of the displayed harmonics.

INTRODUCTION TO SPECTRUM ANALYSIS

OBJECTIVES:

> **1.** To become familiar with the operation of a modern, hardware spectrum analyzer.
>
> **2.** To view the spectral content of sine and square waves.
>
> **3.** To view the spectral content of the amplitude-modulated (AM) sine wave.

REFERENCE:

> Refer to Sections 1-3 and 2-2 in the text.

EQUIPMENT:

> Spectrum analyzer (Rigol DSA-815-TG recommended)
> Function generator or RF signal generator with modulation capability
> Oscilloscope with probe

COMPONENTS:

> 1-kΩ resistor

INTRODUCTION:

> This experiment will introduce you to the operation of a modern, hardware spectrum analyzer similar to one you would encounter on the job. We will look at both sinusoidal and nonsinusoidal signals, as well as an AM sine-wave carrier. The Rigol instrument is specified for this experiment because it is accurate, full-featured, and affordable, and because its user interface is very similar to that of analyzers from other manufacturers. Your instructor will inform you of any needed modifications for use with the analyzers in your laboratory.
>
> As first introduced in Experiment 2, a spectrum analyzer is a test instrument that provides a frequency-domain display of all signal components applied to its input. At its most basic level, the device is a receiver that measures signal levels over a user-selected frequency range. The architecture of a "classic," swept-tuned spectrum analyzer is shown in Figure 3-1.

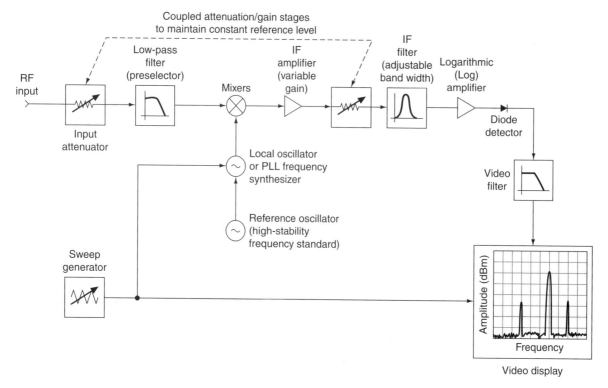

FIGURE 3-1 Block diagram of a swept-tuned, superheterodyne spectrum analyzer.

Its operation is fundamentally very similar to that of any other superheterodyne radio receiver. A mixer stage combines the input with a locally generated signal to produce sum and difference frequencies as well as harmonics of these frequencies. Filter stages subsequent to the mixer select one frequency product, generally the difference frequency, as the *intermediate frequency* (IF), which is then rectified, further processed, perhaps digitized, and finally displayed. A ramp generator produces a voltage sweep that causes a left-to-right movement of an electron beam across a cathode-ray-tube display at the same time the ramp causes a proportional change in the locally generated frequency. In this way, the display represents amplitude as a function of time by using IF amplitude measurements to represent input signal levels indirectly.

Traditional spectrum analyzers rely on analog circuitry to carry out the functions of each stage. Requirements for extreme amplifier linearity and well-characterized frequency responses of attenuators and other stages, as well as for very selective filters with narrow bandwidths, have meant that analyzers capable of operating over wide frequency ranges have traditionally been very expensive. Devices capable of resolving closely spaced signals over frequencies extending into the microwave region could easily cost $25,000 or more, making them, for the most part, specialized laboratory instruments rather than practical troubleshooting tools. Modern spectrum analyzers make extensive use of digital signal processing (DSP) techniques to perform digitally many functions traditionally carried out by analog circuits, among them mixing and filtering, with special-purpose microprocessors carrying out mathematical operations in software. These implementations are possible because the availability of high-performance analog-to-digital (ADC) converters has allowed designers to convert analog signals into digital form much earlier in the signal path than had been possible previously. Effectively, this makes for a *digital IF* architecture in which the analog signal in the IF stages of the traditional instrument is replaced by binary data and processed with DSP-based techniques.

Digital IF implementations coupled with other recent advances in high-speed DSP mean that full-featured analyzers with impressive performance specifications are now available for a fraction of the cost of previous-generation instruments. As an example, the Rigol DSA-815-TG analyzer used in this experiment is capable of operation to a maximum input frequency of 1.5 GHz with a resolution bandwidth (relating to the ability to resolve closely spaced signals) of 100 Hz over the entire range—all for less than $1500. What had been a laboratory curiosity now becomes a practical field troubleshooting instrument as well as an ideal tool for studying the characteristics of electronic signals.

PROCEDURE:

Part I: Introduction to Spectrum Analyzer Operation

Take some time to familiarize yourself with the front panel of the spectrum analyzer. If necessary, consult the user manual. In particular, locate the controls or pushbuttons for entering the center frequency (on the Rigol, press FREQ), the frequency span (SPAN on the Rigol), and amplitude parameters including the reference level and amount of attenuation per division (AMPT). Note also that modern spectrum analyzers and other test instruments make extensive use of "soft keys," which are located immediately to the right of the display and whose functions change depending on the instrument state. On the Rigol instrument, the soft keys have a light gray color, while the FREQ, SPAN, and AMPT keys are dark gray. To illustrate the operation of soft keys, first press the AMPT function key. On the display, the options associated with the soft keys now pertain to the vertical axis of the graph, starting at the top with "Ref Level" and proceeding down through "Input Atten" and "Scale/Div" to "Units" and so on. Changing the function to FREQ causes the soft keys to allow the user to select the center, start, and stop frequencies, as well as other parameters of interest along the x (horizontal) axis of the display.

Locate the connector labeled RF Input (usually a Type N connector in the lower-right corner of the instrument). All signals for measurement will be applied to this input. If your analyzer has a "tracking generator" output option, there will be a second N connector, either in the center or on the left-hand side of the front panel, labeled "TG out." Do not apply signals to the tracking generator output.

1. Write down the following three pieces of information from the front panel label next to the RF input connector:

   Impedance: _____ ohms

   Maximum RF input: + _____ dBm

   Maximum dc voltage input: _____ V

These specifications are critical. Spectrum analyzers are capable of measuring extremely low-level signals (often below −100 dBm) and are easily damaged if the input signal is too large. *Never exceed the maximum signal level specified on the front panel, even for a brief moment.* Sometimes an external in-line attenuator may be required between the output of the device under test and the analyzer input.

2. Determine the maximum input voltage that can be safely applied to the analyzer. Use the values for impedance and maximum RF input determined in the previous step.

   a. Referring to Section 1-2 in the text as necessary, convert the dBm value for maximum RF input to volts and write the result here: _____

   b. With the impedance specified, use the expression $V = \sqrt{PR}$ to determine rms voltage: _____

   c. Multiply the rms voltage by 2.828 to determine the peak-to-peak voltage. _____

3. Make sure that no signal is applied to the RF Input of the analyzer. Power on the instrument and observe its display. Find the following information from the display:

   Reference level in dBm: _____

   Vertical scale factor in dB/div: _____

   Center frequency in MHz: _____

   Span in MHz or GHz: _____

   Resolution bandwidth (RBW) in Hz: _____

The *reference level* is the top horizontal line of the analyzer graph and represents the maximum displayed signal amplitude. The *vertical scale factor* is the reduction (in decibels) in amplitude depicted below the reference level. The *center frequency* denotes the frequency to which the instrument is tuned, and the *span* determines the frequency range displayed on either side of the center frequency. The *resolution bandwidth* determines how closely spaced signals can be in frequency and still be distinguished individually with respect to other signals. All these settings are user-selectable.

4. On the grid below, indicate where you found each reading.

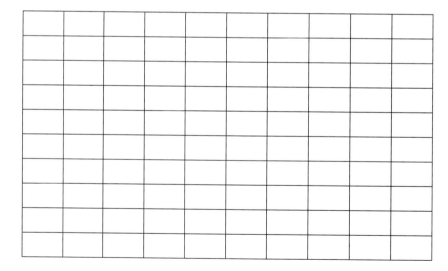

5. To familiarize yourself with instrument operation, use a function generator or RF signal generator to supply a sine-wave signal. Note: Many function generators are capable of supplying signals at amplitudes in excess of the maximum safe amplitude calculated in step 1, possibly leading to equipment damage if high-level signals are applied to the analyzer input. For this reason, it is imperative that, before any signal is applied to the analyzer input, the generator output be set for *minimum* signal output. On many generators, the lowest signal levels are obtained by turning the amplitude control fully counter-clockwise and pulling it out toward you. Additionally, there may be a button labeled −20-dB ATT. Press this button if it is available. *Ask your instructor to verify that the generator is set for minimum amplitude before applying a signal to the analyzer.*

6. With the generator amplitude set for minimum, set its frequency for 500 kHz, sine wave output. Using the appropriate cables and adaptors as needed, apply the signal to the RF input connector of the analyzer and set the analyzer to display the signal. (The following steps apply to the Rigol DSA 815-TG instrument. The front-panel layouts of modern analyzers from other manufacturers are similar; your instructor will supply appropriate modifications to these procedures.):

   a. Set the center frequency: Press the FREQ function key. The top soft key reads "Center Freq." Use the keypad to enter 500 and the third soft key from the top to enter kHz.

   b. Set the span for 10 kHz: Press SPAN, then enter 10 and kHz with the soft keys.

   c. Set the resolution bandwidth: Press BW/Det, then enter 100 and Hz. In the space below, describe what happened to the display:

   d. Change the RBW to 300 Hz and describe what happened:

7. Return the resolution bandwidth to 100 Hz, and, in the grid below, sketch the displayed signal at 500 kHz.

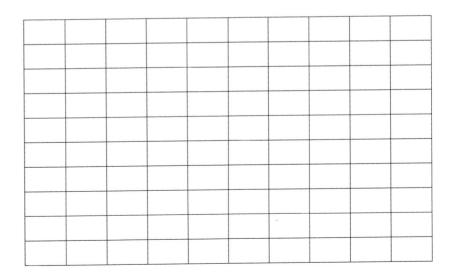

Part II: Spectral Composition of Sine and Square Waves

8. We will now determine the power and amplitude of the sine-wave signal at 500 kHz with the function generator set for minimum amplitude. The Rigol analyzer (like most others) automatically sets the reference level to 0 dBm and the scale/div to 10 dB when it is powered on. Both these settings can be adjusted with the AMPT function key and the appropriate soft keys.

    a. Press AMPT. The Ref Level soft key should be highlighted. Carefully examine the signal at 500 kHz and set the reference level as follows: Determine which dB level is represented by the horizontal line just above where the peak amplitude is shown. For example, if the amplitude falls between −50 and −60 dBm, set the reference level to −50 dBm. Notice that the entire display, including the "noise floor," shifts upward.

    b. Next, change the scale/div to 1 dB with the appropriate soft keys. Record the power level of the 500-kHz signal in dBm here: −_____ dBm.

    c. Convert this level to power in watts. Referring to Section 1-2 in the text as necessary, use the expression

$$10^{(-\text{dBm}/10)} = \frac{P_2}{0.001\,W}$$

to determine the power in watts. Record your result: _____

    d. Using the power determined in the previous step and the system impedance recorded in step 1, determine the amplitude with the expression $V = \sqrt{PR}$ and record it here: _____

9. Return the reference level to 0 dBm and the scale/div to 10 dB. Slowly increase the amplitude of the applied signal to −10 dBm.

10. Press the FREQ function key and set the start frequency to 500 kHz using the Start Freq soft key and the stop frequency to 1.5 MHz using the Stop Freq soft key. The applied signal at 500 kHz, which had previously been in the center of the display, now appears at the far left.

    What is the center frequency now? _____

    Each major division in the horizontal direction represents a frequency change of _____.

    What are the second and third harmonics of the applied signal?____ and ____.

    Are frequency components present at these frequencies? _____

    If so, record the reduction in decibels of the harmonics with respect to the fundamental._____

11. Change the function generator output to produce a 500-kHz square wave. Press BW/Det on the analyzer and the RBW soft key to change the resolution bandwidth to 1 kHz. Press FREQ and set the stop frequency to 3 MHz. Referring to Table 1-3 in the text as necessary, predict the frequencies at which the first two harmonics would be seen: _____ and _____.

    Are the harmonics displayed? _____

    What are their frequencies? _____ and _____

12. Press the start frequency soft key. Now turn the knob counter-clockwise by two or three clicks to shift the start frequency slightly below 500 kHz, thus causing the display to shift to the right. At this point, the fundamental and at least the first two harmonics predicted from step 11 should be visible. Note also that the knob function is determined by the soft key that is active.

13. Record the levels of the fundamental frequency (500 kHz) and of each of the first two displayed harmonics. An easy way to do this is to use the "marker" capability of many modern analyzers to have the instrument read automatically the frequency and amplitude of the highest-amplitude signals within the displayed frequency span. If using the Rigol analyzer, press the white function key marked Peak, located directly above the knob in the front-panel section labeled "Marker." The amplitude and frequency of the fundamental should be displayed. Confirm that the frequency is 500 kHz and record the amplitude below. Then, press the soft key labeled Next Peak. The frequency and amplitude of the first significant harmonic should be displayed. Confirm that the frequency is 1.5 MHz (the third harmonic of the fundamental, and the first significant harmonic of a 500-kHz square wave). Press Next Peak again and confirm that the frequency is 2.5 MHz (the fifth harmonic of the fundamental, and the second significant harmonic). Record the result below:

Fundamental: _____ dBm

first significant harmonic: _____ dBm

second significant harmonic: _____ dBm

By how many decibels is the first displayed harmonic reduced in power from the fundamental? _____ dB. The second displayed harmonic from the first? _____ dB.

Part III: Spectral Composition of Amplitude-Modulated Signals

14. The spectrum analyzer is ideally suited to displaying the frequency components of modulated signals. We can use the setup shown in Figure 3-2 to look at a modulated AM carrier in both the time and frequency domains.

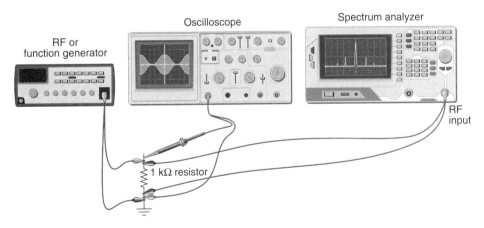

FIGURE 3-2   Test setup for display of amplitude modulated carrier in the time and frequency domains.

The 1-kΩ resistor simply acts as a load and provides a convenient point for measuring the generated signals. Using an RF generator or a function generator with amplitude modulation capability (or the ability to accept a modulating signal from a second generator through an external input connector), apply an unmodulated 500-kHz sine wave at 0 dBm. Referring to Section 2-2 in the text as necessary, predict where the upper and lower side frequencies would be seen if the modulating signal is a 1-kHz sine wave. _____ and _____ kHz.

15. Modulate the carrier with a 1-kHz sine wave. (Note: Most signal or function generators capable of operating at radio frequencies have a built-in modulating signal source. Some generators provide a fixed 1-kHz modulating signal, while others have a variable-frequency modulating signal capability.)

16. Adjust the oscilloscope time base setting for 0.5 ms/div and the vertical sensitivity for 0.1 V/div. At these settings, the 500-kHz carrier will appear as a solid blur covering just over six major divisions vertically. Increase the *modulating signal* amplitude (not the amplitude of the carrier) until the modulation envelope begins to appear. At this point, your oscilloscope display should begin to look like the display depicted in Figure 3-3.

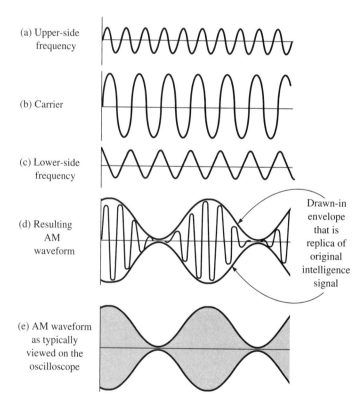

(a) Upper-side frequency

(b) Carrier

(c) Lower-side frequency

(d) Resulting AM waveform

Drawn-in envelope that is replica of original intelligence signal

(e) AM waveform as typically viewed on the oscilloscope

FIGURE 3-3 Carrier and side-frequency components result in AM waveform.

Continue to increase the modulating signal amplitude until the valleys of the modulation envelope just touch each other but do not overlap. At this point, the instantaneous amplitude of the modulated signal at the valleys of the upper and lower modulation envelopes is 0 V. What is the modulation percentage?_____%

17. Set the center frequency of the spectrum analyzer to 500 kHz and the span to 10 kHz. Adjust the Scale/div to 1 dB. Set the resolution bandwidth (RBW) to 1 kHz. How far apart in frequency are the side frequencies above and below the carrier? ____

    Does this result agree with your prediction from step 14 ____? What are the amplitudes of the side frequencies in dBm? ____

    How many decibels below the amplitude of the carrier are the side frequencies?_____

18. Reduce the amplitude of the modulating signal. Does the carrier amplitude change?_____.

19. Return the modulating signal amplitude to the point that produced 100% modulation. Adjust both the amplitude of the signal generator (the modulated signal output) and the vertical sensitivity of the oscilloscope such that the peaks of the modulation envelope extend to eight major divisions vertically. Next, reduce the amplitude of the modulating signal such that the peaks cover six major divisions and the valleys cover two major divisions. This latter point represents 50% modulation. What are the amplitudes (in dBm) of the carrier and of each side frequency?

    Carrier:_____

    Side frequency (upper):_____

    Side frequency (lower):_____

20. From the spectrum analyzer display, what is the bandwidth of the modulated signal?_____

QUESTIONS:

1. Could a digital multimeter (DMM) be used to determine accurately the rms voltage of the 500-kHz signal? Why or why not?

2. In steps 6(c) and 6(d) you changed the resolution bandwidth (RBW) of the instrument from 300 Hz to 100 Hz. Elsewhere, you increased the RBW to 1 kHz. Discuss how the RBW, along with the selected frequency span, affects the appearance of the displayed frequency components as well as the speed at which the display is refreshed.

3. In step 10, would you expect to see harmonics for the 500-kHz sine wave? If harmonics were present, what does this result tell you about the purity of the sine-wave output from the generator?

4. In step 11, are the harmonic frequencies consistent with those predicted by the Fourier series for a square wave?

5. Are the amplitudes of the harmonics for the square wave recorded in step 13 consistent with the Fourier predictions? (Recall that the Fourier series predicts the relative amplitude [voltage] of each frequency component and that power [represented by dBm] scales as the square of voltage.)

6. Does the carrier amplitude vary for an AM carrier as the percentage of modulation is varied? Do the side-frequency amplitudes vary? What do these results tell you about where the information resides in the AM signal?

7. Compare the side-frequency amplitudes in steps 17 and 19. Reducing the modulation percentage from 100% to 50% caused the side frequencies to be reduced in amplitude by how many dB? How much was this reduction in terms of power? What does this result tell you about the importance of maintaining a high average percentage of modulation?

8. In step 20, how did you determine from the spectrum analyzer display the total bandwidth of the AM carrier? How can you determine the frequency of the modulating signal (i.e., the sine-wave intelligence) from the display? Does this result suggest a rule for relating the intelligence frequency to the AM signal bandwidth?

9. Imagine that you are in a job interview. Explain in your own words the similarities and differences between an oscilloscope and a spectrum analyzer. Also explain to the interviewer a typical troubleshooting situation where you would use each instrument.

NAME _____

UPCONVERSION AND DOWNCONVERSION

OBJECTIVE:

To gain an appreciation of signal levels and frequencies for various orders of mixing products, used in upconverters and downconverters.

REFERENCE:

Refer to Sections 2-2, 4-4, and 6-5 in the text.

EQUIPMENT:

Function generator (2)

Spectrum analyzer, with a resolution bandwidth of 3 kHz or less

COMPONENTS:

Mini-Circuits® Model ZP-3 double-balanced mixer

Mini-Circuits® Model ZFL-500HLN 20 dB amplifier

INTRODUCTION:

An *upconverter* uses a mixer and a local oscillator. The "order" of the mixing product is defined as $m + n$, where m and n are integers used in the following equation:

$$F_0 = mF_1 \pm nF_2,$$

where F_0 = output frequency.

F_1 = input frequency.

F_2 = local oscillator frequency.

In order to minimize the conversion loss between frequencies, the local oscillator power usually needs to be at least +5 dBm. In order to avoid distortion, the input frequency needs to be equal to or less than −20 dBm. The theoretical conversion loss when $m + n = 2$ is 6 dB. Consider the following example:

1. $F_2 = 20$ MHz and $F_1 = 1$ MHz.

2. When $n = 1$ and $m = 1$, the "upconverted" frequency would be 21 MHz. If the power at F_1 is −20 dBm, then the theoretical power at 21 MHz would be −26 dBm.

3. Note that the local oscillator power does not affect the output power, as long as it exceeds the minimum level of +5 dBm. For local oscillator power less than +5 dBm, the mixer is "starved" and the conversion loss increases.

4. Note than $m − n$ produces a frequency of −19 MHz. A negative frequency is 180° out of phase with a positive frequency. A typical spectrum analyzer has no way of distinguishing a positive frequency from a negative frequency, so a signal will appear at +19 MHz. The power level will be also about −26 dBm.

The same circuit can be used as a *downconverter*. Consider the following example:

1. $F_2 = 20$ MHz and $F_1 = 21$ MHz. (Note: You can set $R = 3$ MHz and $L = 2$ MHz if your function generators won't go to 20 and 21 MHz.)

2. When $n = 1$ and $m = 1$, the "downconverted" frequency would be 1 MHz. If the power at F_1 is −20 dBm, then the theoretical power at 1 MHz would be −26 dBm.

3. The frequencies for "higher order" modes ($m + n > 2$) can be calculated. These are calculated for all orders, up to 5 and shown in Table 4-1.

4. The approximate conversion loss increases at a rate of 6 dB per order. For example, a third order product has 12 dB conversion loss and a fourth order product has 18 dB conversion loss. Note that these approximations are for a single mixer. The Mini-Circuit ZP-3 is a "double-balanced mixer" so the actual power level of the mixing products may be substantially different from the 6 dB per order approximation.

PROCEDURE:

1. Use a coaxial cable to connect one function generator to the mixer. Set the frequency at 21 MHz and the power level at −2 dBm.

2. Remove the coaxial cable from the spectrum analyzer and reconnect it to the mixer port labeled "R". Refer to Figure 4-1 for the location of "R."

3. Use a coaxial cable to connect the other function generator to the mixer. Set the frequency at 20 MHz and note the maximum power level available. If this power level is less than +5 dBm, use the 20 dB amplifier and set the power level to +10 dBm.

4. Remove the coaxial cable from the spectrum analyzer and reconnect it to the mixer port labeled "L".

5. Use another coaxial cable to connect the spectrum analyzer to the mixer port labeled "X".

6. Measure the power level at all of the frequencies shown in Table 4-1.

TABLE 4-1   Calculated Signals Generated for a Mixer, When $F_2 = 20$ MHz and $F_1 = 21$ MHz (−20 dBm).

#	FREQUENCY (MHz)	m	n	ORDER	* CALC. POWER (dBm)	MEASURED WITH −20 dBm INPUT AT 21 MHz
1	1	1	1	2	−26	
2	2	2	2	4	−38	
3	20	1	0	1	—	
4	21	0	1	1	−20	
5	22	1	2	3	−32	
6	23	2	3	5	−44	
7	19	2	1	3	−32	
8	39	3	1	4	−38	
9	41	1	1	2	−26	
10	43	1	3	4	−38	
11	59	4	1	5	−44	
12	61	2	1	3	−32	
13	62	1	2	3	−32	
14	64	1	4	5	−44	
15	81	3	1	4	−38	
16	82	2	2	4	−38	
17	83	1	3	4	−38	
18	101	4	1	5	−44	
19	102	3	2	5	−44	
20	104	1	4	5	−44	

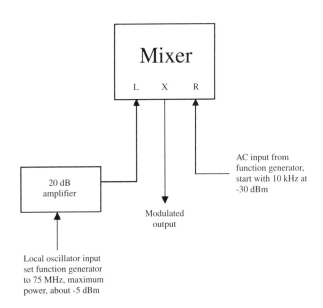

FIGURE 4-1   Mixer Connections

QUESTIONS:

1. From inspection of your results earlier, which frequency combinations (orders) produced the highest-output signals?

2. What are the two possible frequency outputs for the input combination in row 1? These are known as second-order products.

3. From your results, and referring to Section 6-7 in the text as necessary, list the four possible frequencies that can be created from the input frequencies specified for each of the combinations in rows 5 and 7. These are known as third-order products.

4. Again referring to Section 6-7 in the text as necessary, and with your answers to the previous questions in mind, explain why third-order products (and, to a lesser extent, other odd-order products such as fifth-order) are of significance in the design of receivers. Specifically, how do the frequencies predicted from Question 3 affect the receiver's ability to filter out unwanted signals produced as the result of mixing?

5. Why is dynamic range an important consideration in communications receiver design? How do the mixing products predicted in Question 3 affect dynamic range?

6. Must a mixer be an active electronic device? (An active device is one requiring external power to function, such as a transistor or operational amplifier.) List several types of electronic devices that can function as mixers.

7. Could a non-electronic device (such as a rusty fence post) function as a mixer in the presence of a high voltage field? What essential characteristic must any device have for mixing to occur?

FREQUENCY MODULATION: SPECTRUM ANALYSIS

OBJECTIVES:

1. To observe, in the frequency domain, the number and strength of sidebands produced when a sine-wave signal frequency-modulates a carrier.
2. To predict the bandwidth of a transmitted frequency-modulation (FM) waveform using Bessel functions and Carson's rule.
3. To determine the error between predicted and measured results caused by nonlinearities in the modulation stages of signal generating equipment.

REFERENCE:

Refer to Sections 3-2 and 3-3 in the text.

EQUIPMENT:

2 function generators; at least one with external modulation (voltage-controlled frequency [VCF] or voltage-controlled generator [VCG]) capability

2 BNC to BNC cables

BNC tee connector

Oscilloscope probe (×10, 100 MHz)

RF cable (BNC plug to alligator clips)

Spectrum analyzer (Rigol DSA 815-TG recommended)

Frequency counter

dc power supply

COMPONENTS:

None.

INTRODUCTION:

The purpose of this experiment is to demonstrate the spectral characteristics of an FM signal with the spectrum analyzer. The frequency-domain makeup of an FM signal consists of many more components than that of the amplitude-modulated (AM) carrier, even if the modulating signal (intelligence) is a single-frequency sine wave. Whereas a sine-wave-modulated AM wave would always consist of the carrier as well as components at the sum and difference of the carrier and intelligence frequencies and would always have a bandwidth equal to twice the intelligence frequency, a sine-wave-modulated FM carrier would produce a theoretically infinite number of side-frequency components on either side of it, with each one spaced apart from its nearest neighbor by an amount equal to the frequency of the modulating signal. This result implies that FM signals have unlimited bandwidths, rendering FM useless as a means of communication in any bandwidth-limited medium. However, because in general the side-frequency amplitudes get smaller the farther they are from the carrier, only some side frequencies of an FM signal have enough power to be significant. Determining the occupied bandwidth of an FM signal involves identifying the number of significant side frequencies as well as their frequency separation from each other and from the carrier.

A mathematical tool known as the *Bessel function* must be invoked to predict the number of significant side frequencies as well as the amplitude or power of each component. Bessel functions for this purpose have been reduced to tabular form and are standard engineering references. Table 3-1 in the text is a Bessel-function table that shows the number of significant side frequencies and the relative amplitudes in each set of side frequencies as well as in the carrier.

To use the Bessel table, one must determine the modulation index, m_f, of the modulated signal. The modulation index, m_f, is defined as the maximum frequency deviation (that is, the maximum frequency shift caused by the intelligence signal) divided by the modulating frequency:

$$m_f = \text{FM modulation index} = \frac{\delta}{f_i},$$

where δ is the deviation from the carrier (*either* above or below the carrier frequency, not both) and f_i is the modulating frequency. Thus, if a 1-kHz modulating signal at some amplitude causes a 3-kHz deviation above and below the carrier, the modulation index would be 3.0. Note that both the frequency and amplitude of the modulating signal have an effect on the modulation index. (Changes in modulating signal amplitude affect the deviation, thus affecting the number entered in the numerator of the equation for the modulation index.) Because both the deviation and modulating frequency are constantly changing in the presence of a modulating signal consisting of many frequencies, such as voice or music, it follows that the modulation index is also constantly changing. In addition, unlike in an AM system, where the modulation index cannot exceed 1.0 without causing overmodulation, in FM the index can exceed 1.0 and often does.

While Bessel functions are necessary for accurately determining occupied bandwidth and power, an approximation known as *Carson's rule* can be used to predict FM bandwidth. Carson's rule is expressed as follows:

$$BW \approx 2(\delta_{\text{max}} + f_{i_{\text{max}}}),$$

where δ_{max} is the peak deviation from the carrier (either positive or negative), and $f_{i_{\text{max}}}$ is the highest modulating (intelligence) signal frequency. The bandwidth (BW) approximation given by Carson's rule includes about 98% of the total power; the remaining 2% is in the highest-order sidebands outside the predicted bandwidth.

The FM signal is best observed in the frequency domain with a spectrum analyzer. This experiment demonstrates how displays of the carrier and sidebands can be analyzed to confirm the values predicted by the Bessel functions and Carson's rule. The following procedure steps are written around the Rigol DSA-815 spectrum analyzer (frequency range 9 kHz to 1.5 GHz) and low-cost function generators with external modulation feature and a maximum range frequency range greater than 200 kHz. Spectrum analyzers from other manufacturers have a front-panel layout and menu structure very similar to that of the Rigol unit, and the procedure steps can be modified with minimal difficulty. The carrier frequency assumed for the following procedure steps is 200 kHz with an intelligence frequency of 1 kHz. These frequencies are not critical and may need to be modified depending on the maximum frequencies available from the function generators. Your instructor will inform you if modifications (such as a lower carrier frequency) are necessary.

PROCEDURE:

Part I: Determination of Deviation Constant

Two function generators will be used in this experiment, one to produce the carrier and the other to supply the intelligence. Most function generators with an external modulation capability have an input jack (often a BNC connector and often located on the rear panel) that causes the output frequency to shift by some predetermined amount in response to an applied dc or ac voltage. Names vary by manufacturer, but some of the more popular labels for this connector are VCG (voltage-controlled generator) input, VCF (voltage controlled frequency), or FM input. The instantaneous voltage applied to this input determines the deviation in frequency output.

The extent of the frequency change will vary from one generator to another as well as among manufacturers and may also be dependent on the frequency range selected. For this reason, it will first be necessary to determine the deviation constant, k, for the generators in your laboratory for the frequency range of the carrier. Recall from Section 3-2 in the text that the deviation constant defines how much the carrier frequency will deviate (change) for a given input voltage level and has units of kHz/V. As an example, if the generator frequency shifts by 10 kHz for each 1 V of applied voltage, then $k = 10$ kHz/V.

Linearity in the frequency shift is also important, so in the example just given a 1-kHz shift should occur when the applied voltage is 0.1 V (100 mV) and a 5-kHz shift would occur at 0.5 V. The polarity of the applied voltage would determine the direction of the frequency shift; in most cases, a positive applied voltage would increase the output frequency, and a negative voltage would decrease it. We will determine the value of k using the following procedure:

1. Set up the equipment as shown in Figure 5-1.

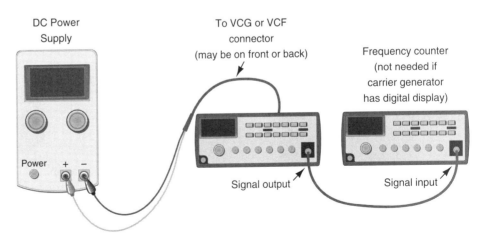

FIGURE 5-1 Equipment setup for determination of deviation constant.

In the figure, the function generator will be used to produce the modulated carrier, and the dc power supply will be connected to the VCG/VCF connector. (The frequency counter would not be needed if the function generator has a digital readout of frequency.)

2. Adjust the function generator output to produce a 500-mV$_{p-p}$ sine wave at 200 kHz with the power-supply voltage initially set to 0 V. Next, increase the power supply voltage until the output frequency has increased by 1 kHz as displayed on the frequency counter. (Note: The dc voltage required to achieve this frequency shift may be small, on the order of 0.1 V.)

3. Reverse the polarity of the power supply leads applied to the VCG input and confirm that the applied dc voltage that caused the generator to *increase* in frequency by 1 kHz in step 2 now causes the frequency to decrease by the same amount. (Simply reverse the leads of the power supply to create a negative voltage at the VCG input.)

4. Return the polarity of the power supply voltage applied to the VCG input to that of step 2 (usually, this means that the positive power supply connection is to the center and negative to the shield of the VCG BNC connector). Return the power supply voltage to the value that caused the output frequency to increase by 1 kHz. Next, increase the power supply voltage by eight equal increments and record the frequency at each increment. For example, if the dc voltage that caused the output frequency to increase by 1 kHz was 0.1 V, then increase the dc voltage to 0.2 V, 0.3 V, 0.4 V, and so on, recording the frequency at each increment.

5. Repeat step 4 using a negative dc voltage applied to the VCG input jack.

6. Plot the graph of the output frequency versus VCG input voltage for your generator. It should be linear or very nearly linear.

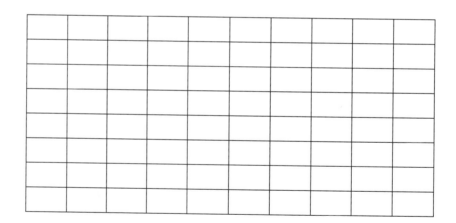

From your graph, determine the deviation constant, k, for your generator by reading off the slope of the straight line or by using the following equation:

$$k = \frac{\Delta f}{\Delta V_{VCG}} = \frac{f_{max} - f_{min}}{V_{max} - V_{min}}$$

$$k = \underline{\hspace{2cm}}$$

Part II: Determination of Reference Carrier Amplitude

Because most function generators do not have calibrated amplitude controls, you will need to use an oscilloscope to determine the precise amplitudes of both the carrier and the intelligence. We will first determine the amplitude necessary to produce a 0-dBm signal at the carrier frequency.

7. Set the function generator to produce a sine-wave output at a frequency of 200 kHz. Using an oscilloscope with the appropriate probe, adjust the generator amplitude until the peak-to-peak voltage is equal to the amplitude for a 0-dBm signal. (Note: There should be no modulating signal or dc voltage applied to the VCG input for this adjustment.) Use the following procedure:

   a. Determine the output impedance of the function generator. This value should be labeled on the front panel. Many generators capable of producing radio frequencies have 50-Ω input impedances, but some audio-only generators have 600-Ω impedances.

   $$Z = \text{_____ ohms}$$

   b. Calculate the rms voltage needed to produce a 0-dBm output. (Remember the definition of 0 dBm and that $P = V^2 \div R$, so rearrange for V and use the impedance determined in the previous step for R.) The result is the root-mean-square (rms) voltage at 0 dBm.

   $$V_{rms} = \text{_____}$$

   c. Multiply the rms voltage you determined in the previous step by 2.828 to obtain the peak-to-peak voltage. Use the oscilloscope to set this amplitude as accurately as possible. After the 0-dBm amplitude has been set, do not touch the amplitude output control again until instructed.

   $$V_{p\text{-}p} = \text{_____}$$

8. Turn on the spectrum analyzer and prepare it to view the function generator output as follows:

   a. Set the center frequency to 200 kHz: For the Rigol instrument, press $\boxed{\textbf{FREQ}}$, $\boxed{\textbf{Center Freq}}$, $\boxed{\textbf{2}}$, $\boxed{\textbf{0}}$, $\boxed{\textbf{0}}$, $\boxed{\textbf{kHz}}$. (Use the number keys on the front panel to input the digits two, zero, and zero.)

   b. Set the span to 10 kHz: Press $\boxed{\textbf{SPAN}}$, $\boxed{\textbf{1}}$, $\boxed{\textbf{0}}$, $\boxed{\textbf{kHz}}$.

   c. Set the resolution bandwidth to 100 Hz: Press $\boxed{\textbf{BW/Det}}$, $\boxed{\textbf{RBW}}$, $\boxed{\textbf{1}}$, $\boxed{\textbf{0}}$, $\boxed{\textbf{0}}$, $\boxed{\textbf{Hz}}$.

   d. From the screen, find the following information and write the information in the spaces below:

   Reference level (dBm): _____

   Vertical scale factor (dB/div): _____

   Center frequency (kHz): _____

   Span (kHz): _____

   Resolution bandwidth (RBW) in Hertz: _____

   e. The *reference level* is the top horizontal line of the analyzer graph and represents the maximum displayed signal amplitude. The *vertical scale factor* (labeled Scale/Div on the Rigol instrument) is the reduction (in decibels) in amplitude depicted by each horizontal line below the reference level. The analyzer defaults to a reference level of 0 dBm when it is turned on. The currently set reference is displayed in the upper left-hand corner of the screen next to the word "Ref." If necessary, set the reference level to 0 dBm by pressing $\boxed{\text{AMPT}}$, $\boxed{\text{Ref Level}}$, $\boxed{\text{0}}$, $\boxed{\text{+dBm}}$. Note that the reference level can be shifted up and down as necessary to view high- or low-amplitude signals. The vertical scale factor defaults to 10 dB/div but can be changed to examine small variations in amplitude between signals.

Again, *without touching the amplitude control of the function generator,* connect the output of the generator to the RF input of the analyzer. A narrow vertical spike should be visible, as seen in Figure 5-2, representing the frequency and amplitude of the signal applied to the RF input.

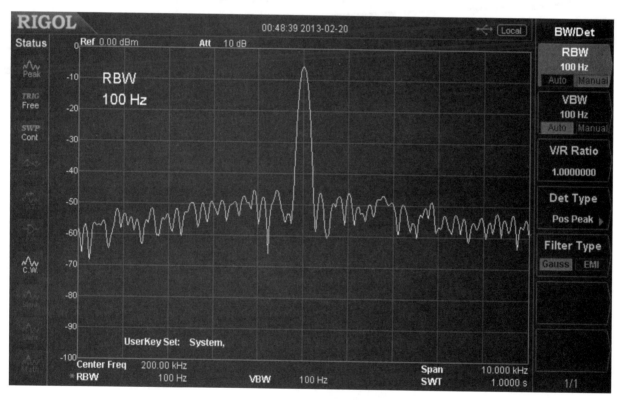

FIGURE 5-2   Spectrum analyzer display of 200-kHz carrier.

The frequency is shown in the horizontal (x) direction, while the amplitude is represented by the height of the spike in the vertical (y) direction. Next, determine the exact amplitude of the signal applied to the analyzer in dBm by changing the vertical scale to 1 dB per division. Press $\boxed{\text{AMPT}}$, $\boxed{\text{Scale/Div}}$, $\boxed{\text{1}}$, $\boxed{\text{dB}}$. Note that the vertical scale is now labeled in 1-dB increments, rather than in 10-dB increments as before. In step 7(c), you set the generator output to 0 dBm. If necessary, disconnect the generator from the analyzer and verify with the oscilloscope that the peak-to-peak voltage of the signal at generator 2 is the same as your calculated value in step 7(c). Then, reconnect the generator to the analyzer. Does the frequency-domain signal displayed show a 0-dBm amplitude? If not, record by how many decibels below 0 dBm the amplitude appears. _____ dB.

9. If necessary, increase the generator signal amplitude to produce 0 dBm as seen on the analyzer. Leave the amplitude control of the generator alone for the remainder of the experiment. Also, adjust the frequency if necessary to re-center the spike on the screen.

10. Change the vertical scale on the analyzer back to 10 dB/div: **AMPT**, **Scale/Div**, **1**, **0**, **dB**.

Part III: FM Signal Analysis

11. Set up the equipment configuration shown in Figure 5-3. The generator whose frequency was set for 200 kHz and whose amplitude was set for 0 dBm in the previous steps will be called the "carrier generator" in this and the following procedure steps, and the second function generator, whose purpose in this experiment is to produce the sine-wave modulating signal, will be referred to as the "intelligence generator." Set the frequency of the intelligence generator to 1 kHz, sine-wave output, and adjust its amplitude to minimum by turning it fully counter clockwise. (Note: On many generators, the lowest signal levels are obtained by pulling the amplitude control out. Also, there may be a button labeled –20-dB ATT. Press this button as well if it is available.) Confirm that the frequency of the intelligence generator is still 1 kHz. Connect the output of the intelligence generator through the tee connector to the VCG input connector of the carrier generator. Using the oscilloscope probe, connect the oscilloscope to the tee connector using a BNC-to-alligator clip coaxial cable as shown in Figure 5-3. The oscilloscope will be used to monitor the peak-to-peak voltage of the modulating signal from the intelligence generator. To review, the output of the intelligence generator is an audio-frequency sine wave whose amplitude is displayed on the oscilloscope at the same time its output is applied to the VCG input of the carrier generator.

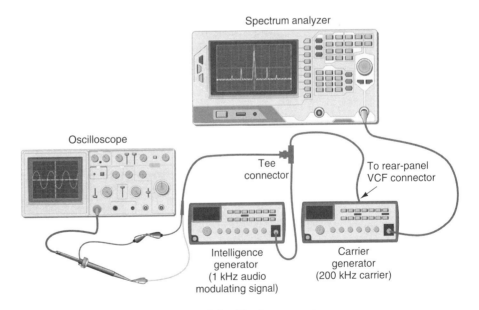

FIGURE 5-3 Equipment setup.

A voltage applied to the VCG input of the carrier generator will cause its output to shift in frequency by an amount proportional to the amplitude of the audio signal, which is the definition of frequency modulation.

12. Plot the display on the spectrum analyzer on the grid below:

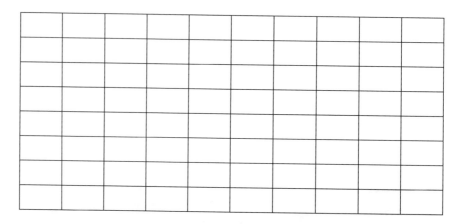

13. Slowly increase the amplitude of the modulating signal from the intelligence generator while viewing the analyzer display. What is happening to the carrier amplitude and number and amplitude of the sidebands as the modulating signal amplitude increases?

14. Use the deviation constant, k, determined in Part 1 to determine the amplitude of the intelligence signal needed to produce a modulation index, m_f, of 1.0. As an example, if you determined that the carrier frequency shifts by 1 kHz for every 100 mV (0.1 V) of voltage applied to the VCF input of the carrier generator, then it follows that +100 mV will cause the carrier frequency to increase by 1 kHz, and −100 mV will cause the frequency to decrease by the same amount. Therefore, for this example, the peak-to-peak amplitude of the modulating signal observed on the oscilloscope would have to be 200 mV to cause a frequency shift of ±1 kHz. With a modulating signal frequency of 1 kHz, the m_f will be 1.0. Plot the displayed signal on the grid below.

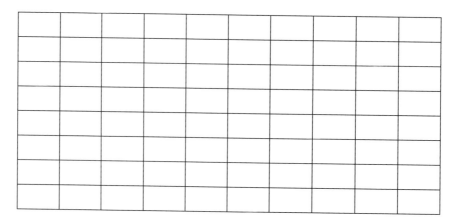

15. On the table of Bessel functions, the carrier is denoted as J_0, while the first *pair* of sidebands (those closest in frequency to the carrier and located above and below it) is denoted as the J_1 pair, the second pair as J_2, and so on. On your sketch earlier, label the carrier as J_0, the first pair as J_1, followed by J_2, J_3, and so forth. How far are the sidebands from each other in frequency? Increase the modulating signal to 1.5 kHz. What happened to the frequency spacing of the sidebands?

16. Return the modulating signal frequency from the intelligence generator to 1 kHz and change the vertical scale factor on the spectrum analyzer back to 1 dB/div, as you did in step 8.

17. Find the J_1 set of sidebands. This will be the set of sidebands closest in frequency to the carrier. From the display, determine by how many dB the amplitude of one of the two sidebands (either upper or lower) has been reduced from the reference level and write it here: $-\_\_$ dB

The table of Bessel functions shown in Table 3-1 in the text expresses the amplitudes of the carrier and sidebands as normalized values, that is, as values referenced to the amplitude of the carrier with no modulation applied. (Normalization is achieved by dividing all the amplitudes by the same reference.) If the unmodulated carrier had an amplitude of 1 V, then the actual voltages for the carrier and sidebands at any modulation index, m_f, would be read directly from the table. For any unmodulated carrier amplitude other than 1 V, the actual voltage for the frequency component of interest (i.e., carrier or sideband) when modulation is applied would be determined by multiplying the decimal fraction shown in the table for that frequency component and modulation index by the voltage of the unmodulated carrier. In step 9, you set the amplitude of the unmodulated carrier to 0 dBm. This amplitude is your reference. You can now compare the measured amplitudes of the modulated carrier and sidebands to the reference value you set. The ratio V_2/V_1 (where V_2 is the amplitude of the modulated carrier or sideband under consideration and V_1 is the 0-dBm reference amplitude) for the carrier or for any sideband in the presence of modulation should equal the decimal fraction for the frequency component shown in the table. Let us see if this is so.

18. First, rearrange the basic expression for voltage ratios in decibel form to solve for the voltage ratio:

$$dB = 20 \log \frac{V_2}{V_1},$$ where "dB" is replaced by the value and sign you recorded in step 17.

$$V_2/V_1 = _____$$

Next, rearrange the previous expression to express the ratio V_2/V_1 in terms of its logarithmic equivalents:

$$\frac{dB}{20} = \log\frac{V_2}{V_1}.$$ (Remember to substitute the number you found in step 17 for "dB" in this and all subsequent steps.)

$$\frac{dB}{20} = _____$$

This expression says that the ratio $dB/20$ is equal to the logarithm of the ratio of the amplitudes of the sideband recorded in step 17 to the amplitude of the carrier with no modulation (the amplitude at 0 dBm you determined in step 9). Remember that logarithms are exponents and that decibels are based on common logarithms (i.e., powers of 10). The common logarithm is the power to which 10 must be raised to equal a given number.

To isolate the fraction V_2/V_1 on the right-hand side of the equals sign, we have to take the antilogarithm of the expression on the left-hand side. The antilog is the inverse of the logarithm—that is, it is the number obtained when the base (10 in the case of common logarithms) is raised to a given power. In our case, the antilog of $dB/20$ is the ratio V_2/V_1:

$$\text{antilog}\,\frac{dB}{20} = \frac{V_2}{V_1}.$$

$$10^{(-dB/20)} = \frac{V_2}{V_1}$$

(On most calculators, the antilog is denoted as 10^x.)

Using the dB value (remember to include the negative sign) you recorded in step 17, determine the ratio V_2/V_1 and record it here:

$$V_2/V_1 = \underline{\hspace{1.5cm}}$$

19. Compare the result you obtained in step 18 with the relative amplitude value for the J_1 set of sidebands for an index of modulation, m_f, of 1.0 shown in the table of Bessel functions. Are the numbers the same? _____

20. Now try a modulation index of 2.0. Ensure that the output from the carrier generator with no modulating signal is still 200 kHz at 0 dBm; adjust as necessary. Leave the frequency of the modulating signal set for 1 kHz. Increase the amplitude of the intelligence generator so that the peak-to-peak voltage as seen on the oscilloscope produces ±2 kHz deviation for the deviation constant determined in part 1.

21. Record in the subsequent spaces the power level in dBm from the spectrum analyzer display for the carrier (J_0) and each indicated sideband. Also, calculate the normalized amplitude for each entry using the procedure outlined in step 18 and show your result in the space to the right of each measured power level.

J_0: _____ dBm; normalized amplitude: _____

J_1: _____ dBm; normalized amplitude: _____

J_2: _____ dBm; normalized amplitude: _____

J_3: _____ dBm: normalized amplitude: _____

22. Return the frequency of the intelligence generator to 1 kHz and increase the amplitude to produce a frequency deviation of ±2.4 kHz and a modulation index, m_f, of 2.4. Carefully adjust the amplitude of the modulating signal up and down until the carrier completely disappears, and record the peak-to-peak voltage of the modulating signal (from the oscilloscope

display) at that point here: _____. Sketch the resulting display below; be sure to label each sideband pair as you did in step 14:

The point where the amplitude of the carrier is zero is known as a *Bessel null* and can be used to calibrate test equipment. The first Bessel null for the carrier occurs at a modulation index of 2.4. From your examination of the attached table, at what modulation indices do the second and third Bessel nulls occur? _____

23. Increase the amplitude of the modulating signal from the intelligence generator and observe what happens to the amplitudes of the carrier and side frequencies. In particular, note the disappearance of the carrier at the second and third Bessel nulls you predicted from step 22. This exercise points out a curious fact of FM: unlike with AM, where the total transmitted power increases in the presence of modulation, with FM the total transmitted power does not change when a carrier is modulated, but the power *distribution* among the carrier and all sets of sidebands is constantly changing as a result of changes of the modulation index. The Bessel table shows relative *amplitudes* of the carrier and sidebands normalized to a value of 1.0. To determine the relative *power* in any signal component, the relative amplitudes must be squared because power increases or decreases as the square of the voltage. Remember that there is only one carrier but there are two sidebands for each *J* value (called the order) created. The power in each sideband (or in the carrier) can be determined by multiplying the relative power for each component by the total power. The sum of each individual sideband power, plus the power in the carrier, should add up to the total power. The complete procedure is shown as Example 3-6 in the text.

24. Carson's rule is a shortcut approximation for determining the occupied bandwidth of a modulated FM carrier. Take the parameters you used in step 14 and predict the bandwidth of the modulated signal using Carson's rule and compare the result with the bandwidth predicted using the Bessel table (see Example 3-7 in the text). How closely do they compare? What would be the effect on the transmitted signal if the bandwidth were reduced to that predicted by Carson's rule?

25. The spectrum analyzer display and Bessel table can be used to determine the modulation index of an FM signal. The normalized voltage ratio (V_2/V_1) of any table entry is converted into decibel form as follows:

$$P_{dB} = 20 \log (V_2/V_1).$$

Remember that each table entry represents the ratio V_2/V_1 for each signal component (that is, carrier or sideband) for a given modulation index. For example, the carrier power at a modulation index of 1.0 is

$$P = 20 \log 0.77 = -2.27 \text{ dB}.$$

For a reference level of 0 dBm, the carrier would, therefore, show a power of -2.27 dBm on the analyzer display.

For a modulation index of 1.5, convert the normalized voltage ratios of the carrier and of each of the first two sideband pairs shown in the Bessel table to their respective power levels. Next, adjust the amplitude of the modulating generator (ensure that its frequency remains at 1 kHz) until the displayed power level of each frequency component matches your calculated power for that component. Then, determine the peak-to-peak amplitude of the modulating signal from the oscilloscope display. Based on the deviation constant for your carrier generator, determine the amount of carrier deviation produced at this amplitude. How closely did the amplitude of the modulating signal match the predicted amplitude for a modulation index of 1.5? Record both the measured and expected amplitudes and determine the error in percentage terms.

Measured amplitude _____

Expected amplitude _____

Error (%) _____

QUESTIONS

1. In step 8, why was the carrier amplitude displayed on the spectrum analyzer approximately 6 dB below the 0-dBm amplitude determined with the oscilloscope?

2. Referring to your spectrum display sketch in step 14 as necessary, describe how you could determine the frequency of the sine-wave intelligence signal from the locations of the side frequencies on the spectrum analyzer display.

3. What is the most likely explanation for any discrepancy between the predicted normalized amplitudes and those you calculated in steps 21 and 22?

4. How closely did the bandwidth determined using Carson's rule (step 24) compare with the result from step 14? What would be the effect on the transmitted signal if the bandwidth were reduced to that predicted by Carson's rule?

5. Describe how the measurements made in step 25 can be used to determine the degree of nonlinearity in an FM modulator. Under what circumstances would knowledge of the Bessel null condition (carrier null) be useful in determining occupied bandwidth in licensed FM systems?

6. Using Example 3-6 in the text as a reference, use the table of Bessel functions to calculate the *power* in the carrier and in each set of sidebands for a 1-kW carrier frequency-modulated by a 2-kHz sine wave. The frequency deviation is ± 3 kHz.

(a) Show the power in the carrier and in each set of sidebands.

(b) Show the total power from your calculations.

(c) Is the total power equal to the power of the unmodulated carrier? What do you think accounts for any discrepancies?

(d) What is the bandwidth of the modulated signal? Is this bandwidth equal to, less than, or more than that of an AM signal with the same modulation?

(e) How does the bandwidth calculation using the Bessel function table compare with the approximation using Carson's rule?

RADIO-FREQUENCY AMPLIFIERS AND FREQUENCY MULTIPLIERS

OBJECTIVES:

1. To investigate the operating principle of a clamping circuit.

2. To study the bias point and operation of a class C amplifier.

3. To examine how a class C amplifier can be used as a frequency multiplier.

REFERENCE:

Refer to Section 4-1 of the text.

EQUIPMENT:

Function generator

Frequency counter with BNC coaxial cable

2 oscilloscope probes (×10, 100-MHz)

Prototype board (breadboard)

COMPONENTS:

2N2222A or 2N3904 NPN transistor

1N914 or 1N4148 signal diode

1 kΩ resistor

120 kΩ resistor

10 mH or 27 mH inductor

Capacitors: 3.3 nF, 1 μF (2), 10 μF (2)

Prototype board (breadboard)

INTRODUCTION:

This experiment demonstrates the basic principles of class C amplifier operation. In Part I, you will investigate how to use a clamping circuit to bias a transistor beyond cutoff, a prerequisite for class C operation. In Part II, you will build a complete class C amplifier with tank circuit, and in Part III, you will see how, by adjusting the resonant frequency of the tank circuit, you can make the amplifier operate as a frequency multiplier.

An amplifier's class of operation is defined by the number of degrees of the input signal over which the active device (transistor or vacuum tube) in the amplifier is turned on. For a class A amplifier, the active device conducts over the full 360° of the input cycle, making the output waveform a faithful reproduction of the input signal. The class B amplifier conducts over 180° of the input signal; such amplifiers must therefore have at least two active devices in a "push-pull" configuration to produce an undistorted output. Class C amplifiers are biased such that the active device is turned on for significantly less than 180° of the input cycle (often 60° or less). Such an amplifier produces brief, high-energy pulses at the collector or plate. When the pulses are applied to a parallel-resonant "tank" circuit, sine waves are produced at the resonant frequency through a phenomenon known as the flywheel effect. The high efficiency and frequency selectivity of class C amplifiers make them very useful in communications circuits.

PROCEDURE:

Part I. Clamping Circuits

1. Build the circuit shown in Figure 6-1. Set channel 1 of the oscilloscope for 0.2 V/div. *Be sure the input coupling is set to dc.* Use channel 1 to monitor the input voltage, V_{in}. Apply a 0.5-V peak-to-peak, 1-kHz sine wave signal to V_{in}. Monitor V_{out} on channel 2 (0.2 V/div, also with dc coupling). Using the vertical position controls for each channel, center the waveforms on the screen so that channel 2 overlaps channel 1. The waveforms at V_{in} and V_{out} should be identical, and when the waveforms are overlaid, the input and output will appear as one waveform.

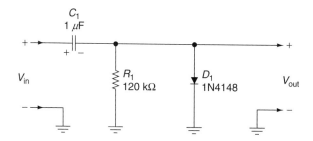

FIGURE 6-1   Clamping circuit for beyond-cutoff bias.

2. Slowly increase the amplitude of V_{in}. Note that the waveform at V_{out} shifts downward with respect to the signal at V_{in}. At the point where the waveforms separate, they are no longer identical with respect to their dc offset.

3. At what V_{in} does V_{out} no longer have the same dc offset? __6.8__ V.

4. Increase V_{in} to 4 V_{p-p} and record V_{in} and V_{out} in Figure 6-2, parts (a) and (b), respectively. Be sure to note in your drawing the peak positive and negative voltages for both waveforms.

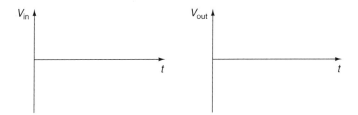

FIGURE 6-2 Input and output waveforms for clamping circuit of Figure 6-1.

5. What happens to V_{out} when V_{in} is increased above the value you recorded in step 3? How does this illustrate the process of negative clamping?

6. Disassemble the circuit of Figure 6-1 and build the circuit of Figure 6-3.

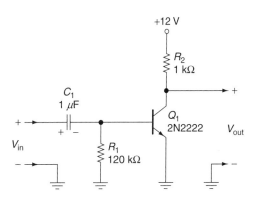

FIGURE 6-3 Transistor clamping circuit.

Note that, in Figure 6-3, the positive side of the dc power supply is connected *only* to the location labeled +12 V at the top of R_2, *not* to the location marked V_{in}. You will apply your sine wave signal to V_{in} as you did in step 1. Both the negative side of the power supply and the black lead of the generator signal cable should be connected to all four ground points shown in the figure. If you are unsure of your wiring, ask the instructor to review your work before applying power.

The diode used in the circuit of Figure 6-1 has been replaced with the base-emitter junction of the transistor. The negative clamping action you observed in steps 3 and 4 will be used to bias the transistor for class C amplification. Class C bias puts the quiescent operating point (Q-point) of the transistor at a point "beyond" cutoff.

a. Again, use channel 1 of the scope (1-V/div, dc-coupled) to monitor V_{in}, and connect channel 2 of the scope (1-V/div, dc-coupled) *to the base of the transistor.* The voltage observed at the base will be designated as V_b.

b. Apply a 1-V_{p-p}, 1-kHz sine wave as V_{in}. Slowly increase the amplitude of V_{in}; you should notice that V_{in} and V_b behave exactly as V_{in} and V_{out} did in step 2.

c. In the graphs shown in Figure 6-4, sketch the V_b waveforms observed when $V_{in} = 1$ V$_{p-p}$, $V_{in} = 2$ V$_{p-p}$, and $V_{in} = 4$ V$_{p-p}$, respectively.

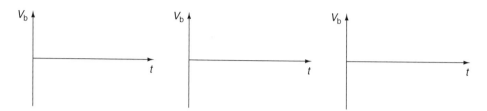

FIGURE 6-4 V_b waveforms observed when $V_{in} = 1$ V_{p-p}, $V_{in} = 2$ V_{p-p}, and $V_{in} = 4$ V_{p-p}.

7. Now move channel 2 of the scope from the base of the transistor to the collector and set channel 2 to 5 V/div with dc coupling. The output waveforms measured at the collector are designated V_{out}. In Figure 6-5, sketch the V_{out} waveforms observed when $V_{in} = 1$ V$_{p-p}$, $V_{in} = 2$ V$_{p-p}$, and $V_{in} = 4$ V$_{p-p}$. Also measure the pulse width of the negative pulse in V_o. From the measurements you made, would you say this amplifier has a gain of less than or greater than unity?

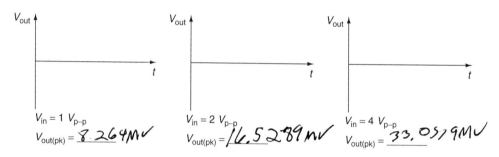

$V_{in} = 1$ V_{p-p}
$V_{out(pk)} = \underline{8.264 mv}$

$V_{in} = 2$ V_{p-p}
$V_{out(pk)} = \underline{16.5289 mv}$

$V_{in} = 4$ V_{p-p}
$V_{out(pk)} = \underline{33.0519 mv}$

FIGURE 6-5 V_{out} waveforms observed when $V_{in} = 1$ V_{p-p}, $V_{in} = 2$ V_{p-p}, and $V_{in} = 4$ V_{p-p}.

From the waveforms you sketched, notice that it takes a certain initial V_{in} voltage level to force the Q-point of the transistor to move into the active region from cutoff. This is what is meant by *beyond cutoff* bias in class C amplifiers.

8. The *efficiency* of this amplifier is fairly high because of the small average level of collector current. Amplifier efficiency, defined as the ratio of useful output power to total input power, is low for class A amplifiers because the active device is operating in its linear region (hence, dissipating power) at all times, even with no signal applied at the input. Class B amplifiers offer higher efficiencies because the active device is cut off for half the input cycle. Operating efficiencies of class C amplifiers are even higher because the active device is biased beyond cutoff and is thus dissipating power for only a small fraction of the input cycle. High operating efficiency is one of the principal advantages of the class C amplifier.

The small average collector current (hence, high efficiency) results from the small duty cycle resulting from negative clamping action as seen in the

V_{out} waveforms of step 6. Calculate the duty cycle for each of the waveforms of step 7 using Figure 6-6 and the following equation:

$$\% \text{ D} = \text{duty cycle} = \frac{\Delta t}{T} \times 100\%.$$

$V_{in} = 1 \text{ V}_{p\text{-}p}; \% \text{ D} = \underline{200 = 20\%}$

$V_{in} = 2 \text{ V}_{p\text{-}p}; \% \text{ D} = \underline{400 = 40\%}$

$V_{in} = 4 \text{ V}_{p\text{-}p}; \% \text{ D} = \underline{800 = 80\%}$

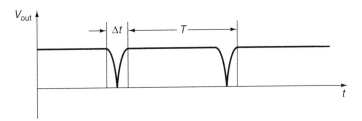

FIGURE 6-6 Calculation of duty cycle from displayed waveform.

Part II. Tuned Class C Amplifier

9. The waveforms of V_{out} observed in step 7 are definitely too distorted to be of any use. Thus, so far, class C amplification appears to be useless. However, the short pulses of V_{out} can be used to activate a parallel resonant circuit. If a short current pulse is applied to a tank circuit, a sinusoidal waveform will be produced as a result of the flywheel effect. Build the class C amplifier circuit of Figure 6-7.

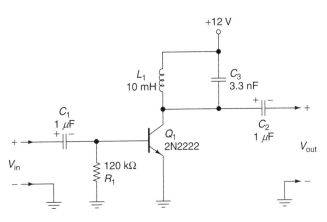

FIGURE 6-7 Class C amplifier circuit.

resonant frequency formula

Monitor V_{in} with channel 1 (1 V/div, dc coupled) and V_{out} with channel 2 (5 V/div, dc coupled). Apply a 1.5-$V_{p\text{-}p}$ sine-wave input voltage set at the resonant frequency of the tank circuit:

$$F_r = \frac{1}{2\pi \sqrt{L_1 C_3}} = \underline{15.91 \text{ kHz}}$$

Fine-tune the frequency of V_{in} so as to produce a maximum value of V_{out}. Fine-tune the amplitude of V_{in} so as to produce an output voltage of 8 V_{p-p}. Sketch the resulting waveforms of V_{in} and V_{out} in Figure 6-8.

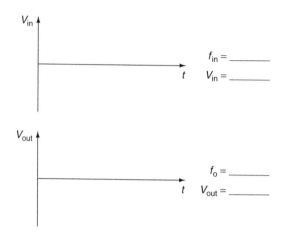

$f_{in} =$ _____
$V_{in} =$ _____

$f_0 =$ _____
$V_{out} =$ _____

FIGURE 6-8   Resulting waveforms of V_{in} and V_{out}.

10. The main limitation of a class C amplifier is the narrow bandwidth of frequencies that can be applied to it. This results from the Q of the tank circuit. Fortunately, in most communications applications a narrow-bandwidth radio-frequency (RF) amplifier is desirable because undesired frequencies can be filtered out. Because of this characteristic, the class C amplifier is sometimes referred to as a "tuned amplifier." Determine the bandwidth of this amplifier by adjusting the frequency of V_{in} above and below the resonant frequency so as to force V_{out} to drop to 70% of the maximum amplitude (8 V_{p-p}) set in step 9. The points above and below the resonant (center) frequency at which the amplitude is 70% of maximum are referred to as the −3 dB points and define the bandwidth of the tuned circuit. The bandwidth is simply the difference between the upper and lower 3-dB frequencies. Record these values here:

$$f_{upper} = \underline{80} \; ; f_{lower} = \underline{20} \; ; BW = \underline{-3}.$$

11. The Q of a tank circuit can be determined using the following equation. Determine this and record your results.

$$Q = \frac{f_r}{BW} \quad Q = \underline{27.7}.$$

Part III. Frequency Multiplication

12. Another use of a tuned class C amplifier is in frequency multiplication. In step 9, notice that for each cycle of the sine wave produced in the tank circuit by the flywheel effect, the tank circuit is recharged by another pulse of collector current. Again, this occurs because of the negative clamping action and the fact that the frequency of the input voltage is exactly matched to the resonant frequency of the tank circuit. However, it is possible to recharge the tank circuit on every other cycle of the sine wave by setting the frequency of V_{in} to exactly one-half of the resonant frequency.

Do this with the circuit you built. Again, fine-tune the frequency of V_{in} so as to produce a maximum value of V_{out}. Also, adjust the amplitude of V_{in} so as to produce $V_{out} = 8$ V_{p-p}. Set the oscilloscope time base and trigger level controls such that multiple cycles of the output waveform are visible (that is, the sine waves do not appear to overlap.) Record the resulting waveforms of V_{in} and V_{out} in Figure 6-9. This is a frequency-doubler circuit.

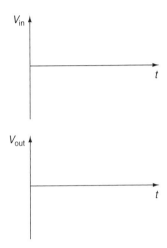

FIGURE 6-9   Resulting waveforms of V_{in} and V_{out} (frequency-doubler circuit).

13. Try adjusting the frequency of V_{in} to one-third, one-fourth, and one-fifth of the input frequency calculated in step 9 so as to produce ×3, ×4, and ×5 frequency multiplication. Measure the input and output frequencies with a counter. Record your resulting input and output voltage amplitudes and frequencies in Table 6-1. In Figure 6-10, record your waveform sketches for the ×5 multiplier.

Note that the voltage amplitude decreases over the later cycles of V_{out} until the next current pulse restores the original amplitude. This distortion can be minimized by the use of higher Q components within the tank circuit. Usually, frequency multiplication greater than 5 is not possible in a single multiplier stage because distortion would be excessive. If larger multiplication constants are needed, more than one stage of multiplication is generally utilized to achieve acceptable results.

TABLE 6-1   Input and Output Voltage Amplitudes and Frequencies.

TYPE	V_{in}	f_{in}	V_o	f_o
×3				
×4				
×5				

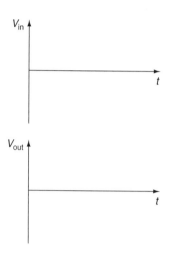

FIGURE 6-10 Waveform sketches for the ×5 multiplier.

QUESTIONS:

1. Would you expect to see the clamping action produced by the circuit of Figures 6-1 and 6-3 if the oscilloscope input coupling were set to ac? Why or why not? *No, Becau*

2. In your own words, and using a load-line diagram if necessary, describe what is meant by biasing a transistor *beyond cutoff.*

3. With reference to Figure 6-7, if the amplitude of V_{in} were varied at an audio rate, would V_{out} also vary in amplitude at the same rate? Why or why not? On the basis of your answer, can you think of a useful application for this amplifier? What is its one principal advantage over class A or B amplifiers?

4. Explain, in your own words, the process of frequency tripling in a ×3 multiplier stage.

5. Would a ×7 multiplier stage work as well as a ×3 multiplier stage? Explain why or why not.

COLPITTS RF OSCILLATOR DESIGN

OBJECTIVES:

    1. To investigate the theory of operation of a Colpitts oscillator.

    2. To follow a "cookbook" design procedure in the fabrication of a working circuit.

    3. To reinforce the concepts of Barkhausen criteria for oscillation.

REFERENCE:

    Refer to Section 4-2 in the text.

EQUIPMENT:

    Dual-trace oscilloscope
    Low-voltage power supply
    Function generator
    Volt-ohmmeter
    Frequency counter
    Prototype board

COMPONENTS:

    2N2222 transistor
    Inductors: 8.2 μH, 27 mH
    Capacitors: 0.1 μF (3), selected design values for C_1 and C_2 from Table 7-1
    Resistors ($\frac{1}{2}$ watt): 680 Ω, selected design values for R_1, R_2, and R_3 from Table 7-1

INTRODUCTION:

    In this experiment you will design, build, and test a Colpitts oscillator. This oscillator belongs to a class of oscillators called resonant oscillators. This name arises from the fact that they use LC resonant circuits as the frequency determining elements. Barkhausen criteria must be met for oscillations to occur. Specifically, the product of the loaded voltage gain of the active device's stage and the attenuation of the feedback network must be equal to or slightly larger than unity to sustain oscillations resulting from an undistorted sinusoidal output signal.

$$A_v \times B \geq 1.$$

Also, the total phase shift that occurs around the closed loop must be close to $0°$ to ensure that positive feedback exists.

Figure 7-1 shows the basic Colpitts oscillator layout. It is easily recognized by the capacitor voltage divider, which makes up the feedback network. Applying Barkhausen criteria to this circuit, we find

$$\frac{V_c}{B_1} \times \frac{V_f}{V_c} \geq 1,$$

where

$$A_v = \frac{V_c}{B_1} \text{ and } B = \frac{V_f}{V_c} = \frac{C_1}{C_2}.$$

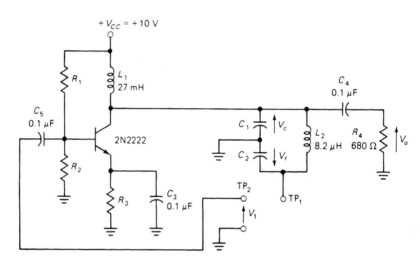

FIGURE 7-1 Colpitts oscillator layout.

In other words, when the circuit is in the open-loop configuration, V_f must be equal to or slightly greater in amplitude to V_1. Also, V_f must be in phase with V_1. If these two conditions are met, then when the loop is closed, the circuit will oscillate. The amplitude criteria are fairly easily met. The voltage-divider action of C_1 and C_2 must yield an attenuation, B, that is the reciprocal of the voltage gain of the transistor amplifier stage, A_v.

$$A_v \geq \frac{1}{B}.$$

Meeting the phase criteria is more difficult. The collector-to-emitter voltage of a common-emitter amplifier is $180°$ out of phase with the base-to-emitter voltage if the collector load is a pure resistance. The feedback scheme of obtaining V_e and V_f in opposite directions with respect to ground causes the additional $180°$ phase shift required to bring V_f to the desired $0°$ phase relationship. T he load on the transistor's collector looks purely resistive when the circuit is resonant. Thus there is no additional phase shift created within the closed-loop system. At resonance, the tank circuit ideally looks like an infinitely large resistance in parallel with R_4. In practice, the equivalent parallel resistance is small enough to decrease the load slightly below 680Ω, because of the Q of the inductor in the tank circuit, so circuit response may be slightly altered from what is ideally expected.

PROCEDURE:

1.2kΩ

RL / (RE + R||prime E)

5.5✓

3.1⁸⁹|⁶⁷²⁹
6✓

4.58 MA

5.45 Ω

Part I: Designing a Colpitts Oscillator Circuit

Design the Colpitts oscillator circuit of Figure 7-1 using the following design procedures. Draw a schematic of the final design in your lab report along with each of your design calculations.

1. *General design rules:* For the final design, use only a single component in each component location. Pick the value of the component from the list of standard value components available in the lab given in Table 7-1. Do this by selecting the standard value that comes closest to each of your design values.

TABLE 7-1 Standard Values of Resistors and Capacitors Available for Use as Design Values.

Resistors (Ω):	10, 15, 22, 27, 33, 47, 68, 82,
	100, 150, 220, 270, 330, 470, 680, 820
	1 k, 1.5 k, 2.2 k, 2.7 k, 3.3 k, 4.7 k, 6.8 k, 8.2 k
	10 k, 15 k, 22 k, 27 k, 33 k, 47 k, 68 k, 82 k,
	100 k, 150 k, 220 k, 270 k, 330 k, 470 k, 680 k, 820 k
Capacitors (nF):	0.1, 0.22, 0.47, 1.0, 2.2, 4.7, 10, 22, 47, 100, 220, 470

2. *Direct current (dc) design:* Determine the values of R_1, R_2, and R_3 to meet the following conditions. Assume that $V_{BE} = 0.5$ V.

   a. The emitter current, I_E, is approximately 3.75 mA.

   b. The current through the voltage-divider resistors, R_1 and R_2, is approximately one-tenth the value of I_E.

   c. $V_{CE} = 5.5$ V. Assume that the 27-mH inductor is ideal (negligible winding resistance).

3. *Alternating current (ac) design:* Assume that the collector load is just R_c. Assume that the tank circuit impedance is too large to produce any noticeable loading on the amplifier stage. Also, assume that the amplifier's input impedance, reflected back through the tank circuit, does not produce any appreciable loading on the output of the amplifier stage.

   a. Determine A_v of the amplifier stage using the values of R_1, R_2, and R_3 determined in the dc design. Use the equations given below to approximate the transistor's base-emitter ac resistance and the voltage gain of the amplifier stage:

$$A_v = -\frac{r_c}{r'_e} = -\frac{R_4}{r'_e},$$

where $r'_e \cong \dfrac{0.025}{I_E}$.

   b. Select single standard values of C_1 and C_2 to achieve a value of B such that $A_v \times B = 10$, and to cause the frequency of oscillation to be 1.8 MHz $\pm$ 200 kHz.

.36
.5

Part II: Colpitts Oscillator Measurements

4. Show the completed design and drawing to the lab instructor for verification that it is complete and accurate. The design and drawing should be completed before coming to the lab, and these are due for inspection at the beginning of the laboratory session.

5. Assemble the circuit on the prototype board. Leave the circuit in the open-loop configuration by leaving the jumper between TP_1, and TP_2 disconnected. Connect the power supply, generator, and oscilloscope to the circuit to make open-circuit measurements.

6. Using the volt-ohmmeter, measure V_{BE}, V_{CE}, and V_E. Determine if the transistor is biased close to the theoretical values used in the design procedures. Document the original data, changes to the circuit design, and final test data.

7. The generator should initially be tuned close to the theoretical resonant frequency of 1.8 MHz. Apply a 20 mV_{p-p} signal at TP_2 and determine the frequency that causes V_c to be of maximum amplitude. Record this frequency. You should use 10:1 probes on the scope to avoid having the scope load down the circuit.

8. At this test frequency, measure the voltage amplitude and frequency of the voltages V_1, V_c, and V_f. Also, record the phase angles of these voltages with respect to V_1.

Record your findings in Table 7-2.

TABLE 7-2   Voltage Amplitude, Frequency of Voltage, and Phase Angles.

VOLTAGE	AMPLITUDE	FREQUENCY	PHASE
V_1			
V_c			
V_f			

9. Using the data gathered in step 8, determine $A_v \times B$ exceeds unity. Show your calculations.

10. If $A_v \times B$ is less than 1, make necessary changes to the circuit in order to cause the product to exceed unity. If the product does exceed unity, close the loop by disconnecting the generator and hooking up a jumper between TP_1, and TP_2. Measure the amplitude, frequency, and phase of the three voltages V_1, V_c, and V_f. Again, make sure that these waveforms are being measured with 10:1 scope probes.

11. Carefully measure and record the waveshape of the oscillator's output voltage. Does it look at all distorted? If it does, consult the troubleshooting chart in Figure 7-2 and make changes to A_v and B to produce the most undistorted output voltage waveform. An easy way to decrease A_v without changing the bias conditions is to place an unbypassed emitter resistance between the emitter resistor and transistor's emitter. Try values between 10 and 100 Ω.

   a. If this resistor is too large, the oscillator will not oscillate.

   b. If this resistor is too small, the output voltage waveform will remain distorted.

104

Symptom #1:   With loop closed, the oscilloscope shows no output signal.

Possible causes:
  (a)   No dc power supplied.
  (b)   1:1 scope probe loading down the circuit.
  (c)   Insufficient voltage gain in amplifier stage.
  (d)   Feedback voltage division too small.

Symptom #2:   With loop closed, the oscilloscope display as shown below:

Possible causes:
  (a)   Way too much voltage gain.
  (b)   Feedback voltage division way too large.

Symptom #3:   With loop closed, the oscilloscope display as shown below:

Possible causes:
  (a)   Slightly too much voltage gain.
  (b)   Feedback voltage division slightly too large.

If the output signal of the oscillator is an undistorted sine wave, then $A_r \times B$ is probably somewhere between 1.0 and 2.0.

FIGURE 7-2   Troubleshooting chart.

.0000 000000000500

2.483647060

$$\frac{1}{4.44}$$

12. Determine a final set of component values that will yield an apparently undistorted sinusoidal signal of desired frequency and record the final schematic in your report.

13. Demonstrate to the lab instructor that your Colpitts oscillator is working properly in producing an undistorted output waveform in the closed-loop configuration.

QUESTIONS:

.225 kHz

1. Draw the ac and dc equivalent circuits for your final working Colpitts oscillator.

2. Explain how your Colpitts design operates, using the results of your measurements and conclusions that you have devised based on positive-feedback concepts and Barkhausen criteria.

HARTLEY RF OSCILLATOR DESIGN

OBJECTIVES:

1. To investigate the theory of operation of a Hartley oscillator.
2. To follow a "cookbook" design procedure in the fabrication of a working circuit.
3. To become familiar with the characteristics of toroid coils and the procedures associated with hand-winding coils.
4. To reinforce the concepts of Barkhausen criteria for oscillation.

REFERENCE:

Refer to Section 4-2 in the text.

EQUIPMENT:

Dual-trace oscilloscope
Low-voltage power supply
Function generator
Volt-ohmmeter
Frequency counter
Prototype board

COMPONENTS:

2N2222 transistor
Capacitors: 0.001 μF, 0.1 μF (3)
Resistors ($\frac{1}{2}$ watt): selected design values for $R_1 - R_5$ from Table 7-1
Toroid core: iron powder type T-106, mix 2 (Palomar or Amidon)
Magnet wire: AWG No. 20 varnish-coated, 4-ft length

INTRODUCTION:

The Hartley oscillator design, shown in Figure 8-1, operates in the same manner as the Colpitts, except that the voltage division is accomplished at the inductive half of the

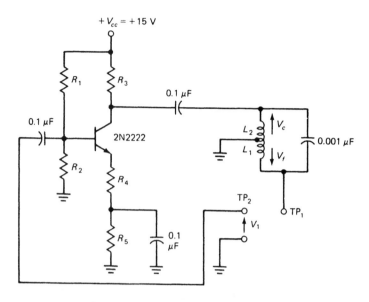

FIGURE 8-1 Tank circuit inductance.

tank circuit instead of at the capacitive half. To make the results agree more closely with theory, the RF choke has been replaced with a fixed collector resistor, R_3. The tank circuit inductance, shown in Figure 8-1 as L_1 and L_2, is fabricated by winding the apropriate number of turns on an iron-powder toroid core. The voltage division is accomplished by connecting a tap at the correct location between the ends of the coil. Iron-powder toroids are circular, doughnut-shaped devices fabricated by bonding very fine iron particles together. They make it possible to build a certain value inductor with fewer turns than would be required without an iron core. They also produce very predictable values of resulting inductance. The toroid shape confines the magnetic field within its circular border and therefore reduces the magnetic coupling from accidentally resulting between coils in neighboring circuits. Winding a single coil on one core simplifies construction and makes the desired voltage division in the feedback network much easier to produce. To find the number of turns needed for type T-106 mix 2 cores, the following relation holds:

$$N = 100\sqrt{\frac{L \times 10^6}{135}},$$

where L is the design value of total inductance in henrys.

The location of the tap can be obtained by applying transformer principles. Since the upper and lower parts of the coil are wound on the same core, the flux cutting each turn is the same. Therefore, the voltage per turn is constant, allowing the following relationship to hold for the feedback attenuation factor:

$$B = \frac{V_f}{V_c} = \frac{N_1}{N_2},$$

where
$\qquad N_1 =$ number of windings for L_1.
$\qquad N_2 =$ number of windings for L_2.
$\qquad N = (N_1 + N_2) =$ total number of windings in the toroid coil.

The major drawback to toroids is that the powdered iron exhibits loss, which appears as a resistor in parallel with the tuned tank circuit. For the T-106 mix 2 variety, the equivalent parallel resistance, R_P, is given by

$$R_P = \frac{K}{N^2},$$

where $K = 619.2 \times 10^3$ Ω-turns^2.
$\qquad$ $N =$ total number of turns.

$\qquad$ This resistance, R_P, is in parallel with R_c in the ac equivalent circuit. The required tap location can be calculated once the voltage-divider ratio, B, is determined. Three turns is about the smallest practical number for obtaining predictable voltage division. When designing the transistor amplifier portion of the oscillator, the best results are usually achieved with gains in the range 5 to 20. Low values of voltage gain create a problem of obtaining the desired voltage division, B, which is very difficult to adjust. High values of voltage gain are hard to obtain repeatedly or predictably from the transistor at high frequencies. They also require the voltage division, B, to be unreasonably small and difficult to achieve.

PROCEDURE:

Part I: Designing the Hartley Oscillator Circuit.

Design the Hartley oscillator circuit of Figure 8-1 using the following design procedures. Draw a schematic of your final design in your lab report along with each of your design calculations.

1. *General design rules:* For the final design, use only a single component in each component location. Pick the value of the component from the list of standard value components available in the lab given in Table 7-1 of the Colpitts oscillator experiment. Do this by selecting the standard value that comes closest to each of your design values.

2. *Alternating current (ac) design:*

$\qquad$ **a.** Determine the value of inductance, L, required to obtain oscillation at 1.8 MHz ± 200 kHz.

$\qquad$ **b.** Determine the total number of turns required to achieve this total inductance.

$\qquad$ **c.** Select a tap location that will yield a value of B somewhere in the range 0.18 to 0.25.

$\qquad$ **d.** Calculate the equivalent parallel resistance, R_P, of the coil at resonance.

$\qquad$ **e.** Calculate the value of voltage gain, A_v, of the amplifier to create $A_v \times B$ to be between 1.0 and 1.1.

$\qquad$ **f.** Select a trial value of R_4 so that R_4 is approximately 15 times greater than the transistor's base-emitter dynamic resistance, r_e'. Use a dc emitter current value of 3.75 mA. Approximate r_e' using

$$r_e' \cong \frac{0.025}{I_E}.$$

$\qquad$ **g.** Calculate the value of ac collector resistance, r_e, required to yield the desired A_v, using

$$A_v = \frac{r_e}{r_e' + r_e} = \frac{R_P \| R_3}{r_e' + R_4}.$$

h. Using the r_c found in step (g) and the equivalent parallel resistance, R_p, found in step (d), calculate a trial value of R_3.

$$r_c = \frac{R_3 R_P}{R_3 + R_P}.$$

3. *Direct current (dc) design:*

    a. Select values of R_1, R_2, and R_5, which when used with R_3 and R_4 from the ac design yields the following dc conditions. Assume that $V_{BE} = 0.5$ V.

        (1) dc emitter current: $I_E = 3.75$ mA.

        (2) $V_{CE} = 6.0$ V.

        (3) The current through the voltage-divider resistors, R_1 and R_2, is approximately one-tenth the value of I_E.

    b. Make sure that R_5 is set at least 10 times as large as R_4.

Part II: Hartley Oscillator Construction and Test.

4. Show the completed design and drawing to the lab instructor for verification that they are complete and acceptable.

5. Assemble the oscillator circuit using the prototype board. Set up the circuit in the open-loop configuration by leaving TP_1 and TP_2 disconnected.

6. Using the volt-ohmmeter, measure V_{BE}, V_{CE}, and V_E, to determine if the transistor is biased properly for amplification to occur. Document the original data, any changes made to the circuit design, and final dc data.

7. The generator should be initially tuned close to the theoretical resonant frequency of 1.8 MHz. Apply a 200-mV$_{p\text{-}p}$ signal at TP_2 and determine the frequency at which V_c is of maximum amplitude. Use 10:1 scope probes to make sure that the scope's internal impedance does not alter the waveforms being observed due to loading.

8. At this frequency, measure the voltage amplitude, frequency, and phase of V_c, V_f, and V_1, using V_1 as phase reference.

9. Using the data gathered in step 8, decide if $A_v \times B$ is greater than unity.

10. If $A_v \times B$ is less than 1, make necessary changes to the circuit design to cause the product to exceed unity and repeat steps 2–6. If $A_v \times B$ is greater than 1, close the loop by connecting a jumper between TP_1 and TP_2 and disconnect the generator from the circuit. Measure the amplitude, frequency, and phase angle of the three voltages as was done in step 5. Again, use 10:1 scope probes.

11. Carefully measure and record the waveshape of the oscillator's output voltage. If the output waveform is not a clean sine wave, but shows distortion, or if there is no output waveform present, consult the troubleshooting chart in Figure 7-2 from Experiment 7. Make appropriate modifications. Record all changes made to your design. An easy way to change the voltage gain of the amplifier stage without changing the bias conditions or the resonant frequency is to alter the value of R_4.

12. Determine a final set of values for the circuit that will yield an apparently undistorted sinusoidal signal of the desired frequency.

13. Demonstrate to the lab instructor that your Hartley oscillator is working properly in producing an undistorted output waveform in the closed-loop configuration.

QUESTIONS:

1. Draw the ac and dc equivalent circuits for the oscillator stage.

2. Explain how your circuit operates, using the results of your measurements and conclusions you have devised based on positive-feedback concepts and Barkhausen criteria.

NAME _____

PRINCIPLES OF NONLINEAR MIXING

OBJECTIVES:

    1. To characterize the properties of a ceramic bandpass filter.

    2. To review concepts of attenuation and insertion loss using decibel units.

    3. To become acquainted with the effects of simultaneously applying voltages of different frequencies to a nonlinear device.

REFERENCE:

    Refer to Sections 4-3 and 4-4 in the text.

EQUIPMENT:

    2 signal or function generators capable of operation in the 455-kHz frequency range with leads

    Frequency counter and cable

    Dual-trace oscilloscope with probes ($\times$10)

    Prototype board (breadboard)

COMPONENTS:

    455-kHz ceramic bandpass filter: Murata CFWLB455KGFA-B0 recommended

    1 each 1-kΩ resistor

    4 each 2-kΩ resistors

INTRODUCTION:

    This experiment further illustrates the mixing process that occurs when two or more ac voltages of different frequencies are applied simultaneously to a nonlinear device. If the applied voltages, denoted f_1 and f_2, are sinusoidal, the nonlinear mixing action will produce several output frequencies, which can be shown mathematically to include the following:

    1. The original frequencies, f_1 and f_2.

    2. The second harmonics of the original frequencies, $2f_1$ and $2f_2$.

3. The sum of the two original frequencies, $f_1 + f_2$.

4. The difference between the two original frequencies, either $f_1 - f_2$ or $f_2 - f_1$, whichever produces a positive result.

5. 0 Hz.

It is important to recognize that mixing only occurs within devices that exhibit *nonlinear* characteristics, such as diodes or transistors. If f_1 and f_2 were applied to the input of a linear device such as a resistor, the result would be simple linear addition of the inputs; that is, both frequencies would be present at the output, but no interaction between the input frequencies would take place and no additional frequency products would be produced.

Mixing is not always desirable. Since the potential for mixing exists whenever two or more signals are combined within a nonlinear device, and because nonlinear devices are used in virtually all electronic circuits, it follows that mixing can occur in any circuit, even if the circuit does not normally function as a mixer (i.e., is intended to be linear). For example, a normally linear amplifier driven into nonlinear operation by the presence of a large, undesired signal will produce spurious frequency products. These undesired products resulting from mixing action can be very difficult to eliminate; indeed, many challenges in communication system design and maintenance involve, at least in part, the prevention or elimination of unwanted mixing products.

In this assignment, you will demonstrate mixing by feeding the complex output signal of a mixer (the diode) through a sharp bandpass filter tuned to 455 kHz. The first part of the assignment will be to verify the characteristics of the filter. The second part will involve selecting various input frequencies and predicting and then measuring the outputs. The two input frequencies, f_1 and f_2, will be carefully selected so that only one of the predicted output frequencies resulting from nonlinear mixing ends up being near 455 kHz, the pass frequency of our filter. The purpose of the filter is to isolate 455 kHz from all the other frequencies produced by nonlinear mixing.

PROCEDURE:

Part I: Bandpass Filter Characterization

1. Carefully examine Figure 9-1.

    This figure shows the narrow band of frequencies around 455 kHz that are passed with minimal attenuation (called the *passband*), as well as frequencies farther away that are substantially reduced in amplitude (called the *stopband*). From the figure, *estimate* the following:

    a. The frequencies above and below 455 kHz where the output voltage would be one-half the input.

    Lower frequency:_____ Upper frequency: _____

    b. The frequencies above and below 455 kHz where the output voltage is attenuated to one-tenth the input.

    Lower frequency: _____ Upper frequency: _____

    c. The frequencies above and below 455 kHz where the output voltage is attenuated to 1/100 the input.

    Lower frequency: _____ Upper frequency: _____

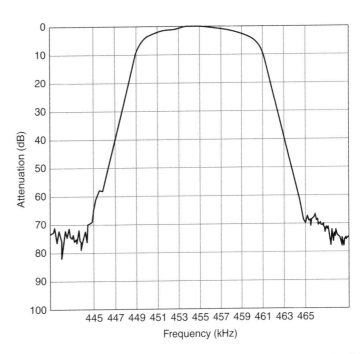

FIGURE 9-1 Frequency response curve for Murata CFWLB455KGFA-B0 bandpass filter.

d. The frequencies above and below 455 kHz where the output voltage is attenuated by at least 45 dB from the passband.

Lower frequency: _____ Upper frequency: _____

2. Construct the test circuit in Figure 9-2.

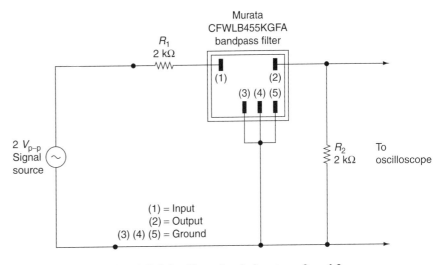

FIGURE 9-2 Test circuit for steps 2 and 3.

Use the function generator to apply a 2-V_{p-p} sine-wave signal at 455 kHz to R_1 and measure the peak-to-peak voltage of the signal *at pin 1 of the ceramic filter.* (That is, measure the voltage directly at the input to the filter, not the voltage applied to R_1 from the generator.) This voltage will be called V_{in}; record V_{in} in the space below. (Note that V_{in} is less than 2-V_{p-p} because of the presence of R_1.) Next, record the peak-to-peak voltage seen across R_2; this is the filter output voltage and will be called V_{out}. Record V_{out} below:

$V_{in} =$ _____V_{p-p}.

$V_{out} =$ _____V_{p-p}.

Was V_{out}, the voltage across R_2, less than V_{in}? Explain why this is so.

Attenuation caused by the presence of the filter in the circuit is known as *insertion loss*. Using the formula $20 \log (V_{out} \div V_{in})$, determine the insertion loss in decibels for your filter at 455 kHz. Compare your measured insertion loss with the specification for your filter (available from the manufacturer's website).

Insertion loss: _____ dB

Spec. met? Yes/No

3. Verify that the input amplitude to R_1 is 2-V_{p-p} and that V_{in} is as measured in step 2 for each of the following steps.

a. Reduce the generator frequency until the output measured across R_2 is 3 dB below the output voltage, V_{out}, measured at 455 kHz and recorded in step 2. Record this frequency.

$f_{lower\ -3\ dB} =$ _____ kHz.

b. Reduce the frequency until the output is 6 dB below the V_{out} voltage measured at 455 kHz and record this frequency.

$f_{lower\ -6\ dB} =$ _____ kHz.

c. Reduce the frequency until the output is 12 dB below the V_{out} voltage measured at 455 kHz and record the frequency.

$f_{lower\ -12\ dB} =$ _____ kHz.

d. Reduce the frequency until the output is 20 dB below the V_{out} voltage measured at 455 kHz and record the frequency.

$f_{lower\ -20\ dB} =$ _____ kHz.

e. Reduce the frequency until the output is 40 dB below the V_{out} voltage measured at 455 kHz and record the frequency (Note: The voltage will be very small and may have to be estimated).

$f_{lower\ -40\ dB} =$ _____ kHz.

f. Repeat steps (a) through (e) for the frequencies above 455 kHz and record the frequencies.

$f_{upper\ -3\ dB} =$ _____ kHz.
$f_{upper\ -6\ dB} =$ _____ kHz.

$$f_{\text{upper}-12\,\text{dB}} = \underline{\hspace{1cm}} \text{ kHz}$$

$$f_{\text{upper}-20\,\text{dB}} = \underline{\hspace{1cm}} \text{ kHz}$$

$$f_{\text{upper}-40\,\text{dB}} = \underline{\hspace{1cm}} \text{ kHz}$$

4. On a separate sheet of paper and using the data you collected from the previous step, construct a frequency-response graph for your bandpass filter. Your graph should show frequency on the horizontal (X) axis and attenuation *in decibels* on the vertical (Y) axis. Make any additional measurements necessary to produce enough data points to produce a graph similar to that of Figure 9-1.

Part II: The Mixer

5. Build the mixer stage shown in Figure 9-3. The nonlinear current-versus-voltage (I-V) characteristics of the germanium diode make it function as a nonlinear mixer.

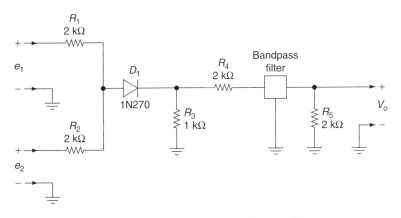

FIGURE 9-3   Mixer stage for Part II.

6. Set the amplitudes of input 1 (e_1) and input 2 (e_2) at 5 $V_{\text{p-p}}$. Set the frequency of e_1 to 455 kHz and the frequency of e_2 to 200 kHz. Monitor the output voltage using the oscilloscope. Carefully fine-tune the frequency of one of the two input generators so that the output voltage reaches maximum amplitude. Once the amplitude is peaked, measure and record the output voltage and frequency.

Voltage: _____ V

Frequency: _____ kHz

7. Repeat step 6 for each of the following input frequencies:

    a. $f_1 = 200$ kHz and $f_2 = 455$ kHz.   V: _____ V   F: _____ kHz

    b. $f_1 = 227.5$ kHz and $f_2 = 300$ kHz. V: _____ V   F: _____ kHz

    c. $f_1 = 300$ kHz and $f_2 = 227.5$ kHz. V: _____ V   F: _____ kHz

    d. $f_1 = 355$ kHz and $f_2 = 100$ kHz.   V: _____ V   F: _____ kHz

    e. $f_1 = 755$ kHz and $f_2 = 300$ kHz.   V: _____ V   F: _____ kHz

    f. $f_1 = 295$ kHz and $f_2 = 80$ kHz.   V: _____ V   F: _____ kHz

QUESTIONS:

1. In Part I, you experimentally characterized the performance of your band-pass filter. Discuss a situation in which such measurements would have to be performed in a production environment. How could these measurements be speeded up and automated to allow for the characterization of a large number of parts in a short period of time?

2. If step 6 were completed using $f_1 = 682.5$ kHz and $f_2 = 227.5$ kHz, what problem would occur in looking for the output difference frequency component of $(f_1 - f_2) = 455$ kHz?

3. In step 7, which outputs— f_1, f_2, $2 f_1$, $2 f_2$, $(f_1 + f_2)$, $(f_1 - f_2)$—were of the largest amplitude? What does this result say about the relative amplitudes of the fundamental frequencies (i.e., f_1 and f_2) and their harmonics? (Hint: look at the discussion of frequency spectra and Fourier transforms in Chapter 1 of the text.)

4. In step 7, the only output frequency that would be within the bandpass of the filter would be $(f_1 + 2 f_2)$, which is one of the smaller, usually ignored, output frequencies. How many decibels down is this output frequency compared to the simple sum and difference frequency outputs that are produced by nonlinear mixing action? How can these "extra" nonlinear modulation products be kept to a minimum amplitude?

5. The frequency component $(f_1 + 2f_2)$ seen in step 7(f) is called a *third-order* frequency component because the sum of the coefficients of the expression f_1 and $2f_2$ is 3. If you apply two frequencies, f_1 and f_2, to a nonlinear device, a third-order product would be created whenever one frequency is mixed with the second harmonic of the other because the sum of the coefficients would be three. (The *mixing* may be either additive or subtractive; the *coefficients*, however, are always additive.) What three other third-order components could result from nonlinear mixing?

6. Based on your answer to question 5, explain why third-order frequency components are of particular importance to communications system designers. (Hint: Remember that all third-order frequency components are undesired, so compare the frequencies you calculated in step 5, which are undesired frequencies, with the desired frequency [455 kHz in this case], and consider the characteristics of the bandpass filter you graphed in Part I of this experiment.)

SIDEBAND MODULATION AND DETECTION

OBJECTIVES:

    1. To become familiar with the 1496 balanced mixer.

    2. To build and evaluate a balanced modulator that produces a double-sideband suppressed carrier signal.

    3. To build and evaluate a product detector that extracts the intelligence from a single-sideband suppressed carrier signal.

REFERENCE:

    Refer to Sections 2-3, 4-4, 5-3, 6-3, and 6-5 in the text.

EQUIPMENT:

    Dual-trace oscilloscope: must have a Y vs X display capability

    Function generator (2)

    Low-voltage power supply (2)

    Frequency counter

    Spectrum analyzer (if available)

COMPONENTS:

    1496P integrated circuit (2)

    Murata CFWLB455KGFA-B0 ceramic filter

    Capacitors: 0.005 μF (3), 0.1 μF (4), 1.0 μF (2)

    Resistors ($\frac{1}{2}$ watt): 47 Ω (5), 100 Ω (2), 1 kΩ (8), 3.3 kΩ (4), 6.8 kΩ (2), 10 kΩ (2)

    Potentiometer: 10 kΩ (10-turn trim)

A balanced modulator (Figure 10-1) is typically used to generate a double-sideband suppressed carrier signal in an SSB transmitter that uses the filter method of design. Nonlinear amplification causes the creation of first and second harmonics of the intelligence input signal, the simple sum and difference frequency components, and a dc component. The first and second harmonics of the carrier input signal, which are normally produced

71

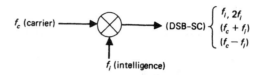

FIGURE 10-1 Balanced modulator function.

in mixing action are suppressed in a balanced modulator. In the 1496, this is done by signal cancellation due to the symmetrical arrangement of the differential amplifier stage as shown in Figure 10-2. A spectrum diagram of the output signal reveals the presence of the lower and upper sidebands but no RF carrier frequency component (Figure 10-3).

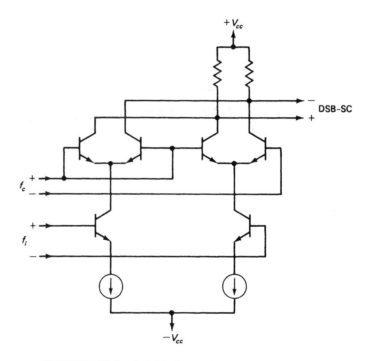

FIGURE 10-2 1496 balanced modulator circuitry.

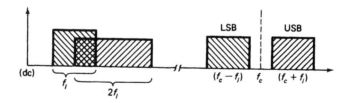

FIGURE 10-3 Balanced modulator typical output spectrum.

To produce single sideband, a sharp filter that exhibits a fairly constant passband response and steep roll-off skirts is needed to attenuate all frequencies produced by a balanced modulator except for the desired sideband. A ceramic filter such as the one used in Experiment 9 may fit these design requirements.

A balanced modulator followed by a low-pass filter is typically used to recreate the original intelligence signal in an SSB receiver. Again, the balanced mixer's nonlinear amplification causes the creation of the first and second harmonics of the input SSB signal, the simple sum and difference frequencies of the two input signals, and a dc signal. Again, the first and second harmonics of the input carrier signal are suppressed by the balanced modulator. This time, however, since the frequencies of the SSB input and carrier input signals are fairly close to each other, the difference frequency components end up being much smaller than any of the other frequencies produced by mixing action. These difference frequencies are exactly equal to the original intelligence frequencies. For example, if upper sideband is being supplied to the SSB input of the balanced modulator, the frequency spectra shown in Figure 10-4 will result in the output signal.

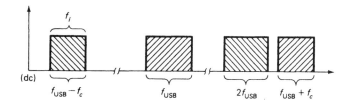

FIGURE 10-4   SSB detector output spectra before filtering.

It is very easy to filter out the original intelligence from this complex signal because of the large frequency difference between the frequencies and all other output frequency components produced by mixing action. A similar result occurs if lower sideband is being supplied to the balanced modulator. All that is necessary is a simple *RC* low-pass filter that has a sufficiently large time constant at all RF frequencies. The balanced modulator, low-pass filter, and RF carrier oscillator make up what is known as a product detector in an SSB receiver.

PROCEDURE:

1. Using the theory described above and in your text, draw the spectrum diagrams of the output signals that would result in each of the following designs.

   a. The output double-sideband suppressed carrier signal if a 475-kHz sinusoidal RF carrier is mixed with a 20-kHz audio sine-wave signal in a balanced modulator.

   b. The output double-sideband suppressed carrier signal if a 435-kHz sinusoidal RF carrier is mixed with a 20-kHz audio sine-wave signal in a balanced modulator.

   c. The output signal if a USB signal having a frequency range of 455 kHz–460 kHz is mixed with a 460-kHz sinusoidal RF carrier in a balanced modulator.

   d. The output signal if an LSB signal having a frequency range of 450 kHz–455 kHz is mixed with a 450 kHz sinusoidal RF carrier in a balanced modulator.

   e. If the output signal of either design (c) or (d) above is passed through a low-pass filter that totally attenuates all signals above 100 kHz but passes all frequencies below 50 kHz, sketch the resulting output spectra of the low-pass filter.

2. Build the balanced modulator circuit shown in Figure 10-5. Apply ±10 V dc to the circuit. Apply a 2 $V_{p\text{-}p}$, 400-kHz sine wave at TP$_1$. This signal represents the input RF carrier. Monitor V_o at TP$_3$ with the oscilloscope. Adjust the carrier null potentiometer, R_2, for a minimum-amplitude 400-kHz sine wave. At the optimum precise setting it should null the carrier to zero amplitude and increase the carrier on either side of the null setting.

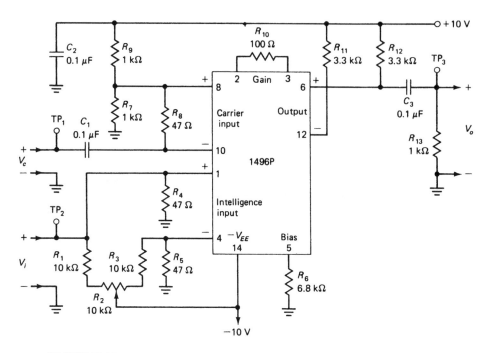

FIGURE 10-5 DSB-SC generator using a 1496P balanced modulator.

3. Apply a 150-mV$_{p\text{-}p}$ 200-Hz sine wave at TP$_2$. This signal represents the input audio intelligence signal. Again, monitor TP$_3$ with the oscilloscope. Use 10:1 probes to avoid loading. The output signal at TP$_3$ should look like a "fuzzy" 400-kHz sine wave if the signal is being viewed with a small horizontal time scale, such as 0.5 μs/div, and is being triggered by the RF carrier signal. Sketch the observed signal at TP$_3$.

4. A better, more informative waveform can be displayed by placing the horizontal time scale at a much larger setting, such as 0.5 ms/div and by externally triggering the scope with the audio intelligence signal. Do this by observing the signal at TP$_3$ with channel A and the signal at TP$_2$ with channel B. The observed DSB-SC signal will probably be slightly distorted. Make small adjustments to the carrier null potentiometer, R_2, and to the intelligence and carrier amplitudes in order to produce a clean waveform such as that shown in Figure 10-6. The two interwoven envelopes should be near-perfect sine waves that have a frequency of 200 Hz. The peak envelope voltages should all be equal to one another. Sketch the resulting waveform.

5. Determine the effect on the resulting DSB-SC waveform if changes are made in the frequencies of the carrier signal or the intelligence signal. Also, determine the effect of changing the gain resistance, R_{10}, to 47 Ω and to 1 kΩ. Return the display back to its original form as in step 4 before proceeding.

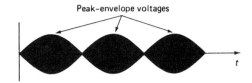

Peak–envelope voltages

FIGURE 10-6 DSB-SC waveform showing no distortion.

6. Observe and sketch the resulting swept-frequency display of the DSB-SC signal by applying the DSB-SC signal to the vertical input of the scope and applying the intelligence signal to the horizontal input of the scope. Place the scope in A versus B mode. Set the vertical and horizontal sensitivities to fill the screen with the display. You should note a double triangle or "bow-tie" shaped swept display. Notice the effect on the screen of changing the amplitude of the carrier and intelligence signals. Notice that when distortion occurs, the two triangles will no longer have straight sides. Try adjusting the value of the carrier null potentiometer. You should notice that the two triangles are symmetrical when the carrier has been nulled. Again, return the display to its original form as in step 4 before proceeding.

7. The double-sideband suppressed carrier signal will now be applied to a ceramic filter to produce true SSB. Recall from Experiment 9 that the Murata CFWLB455KGFA-B0 has a 3-dB bandwidth of approximately 20 kHz and a center frequency of 455 kHz. Adjust the intelligence signal frequency to exactly 20 kHz and the RF carrier frequency to exactly 435 kHz. The DSB-SC signal should now have an upper sideband frequency of 455 kHz and a lower sideband frequency of 415 kHz. Build the circuit in Figure 10-7 and connect it to the output of the balanced modulator circuit by connecting TP_3 and TP_4 together. Monitor the output voltage at TP_5. It should be a clean sine wave, since only the upper sideband frequency component makes it through the filter. Adjust the carrier frequency so that the voltage at TP_5 is at maximum amplitude. Measure its frequency with a counter.

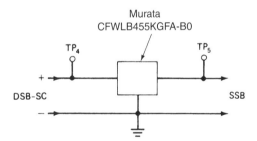

FIGURE 10-7 Ceramic filter to create SSB from DSB-SC.

If a spectrum analyzer is available, observe the spectral content of the signals at TP_3 and TP_5. Sketch the spectral displays.

8. Repeat step 7, but this time allow only the lower sideband frequency component of the DSB-SC signal to make it through the ceramic filter. Since you cannot change the bandpass frequency range of the ceramic filter, you must change the frequency of the RF carrier signal entering the balanced modulator to align the lower sideband frequency with the bandpass.

$$f_{LSB} = \underline{\hspace{2cm}} \qquad f_{carrier} = \underline{\hspace{2cm}}$$

If a spectrum analyzer is available, observe the spectral content of the signals at TP_3 and TP_5. Sketch the spectral displays. Return to the scope display of step 7 before continuing.

9. The SSB signal produced at the output of the ceramic filter in steps 7 and 8 will now be applied to a balanced modulator. Build the circuit shown in Figure 10-8. Connect the output of the ceramic filter to the detector's input by connecting TP_5 and TP_6 together with a jumper. The RF carrier driving the balanced modulator at TP_1 should also be connected to the RF carrier input of the detector at TP_7. Adjust the amplitude of V_c to approximately 3.5 V_{p-p}. Observe the output waveform of the product detector at TP_8 with channel A of the oscilloscope. Monitor the original intelligence signal with channel B of the oscilloscope. Use 10:1 probes.

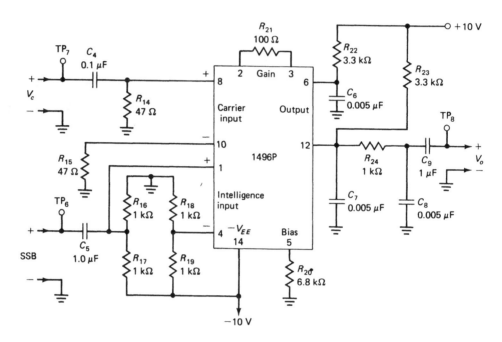

FIGURE 10-8   Product detector using a 1496P balanced modulator.

10. Readjust the frequency of the intelligence for 200 Hz. Adjust the RF carrier frequency so that the detected output frequency signal is a maximum. You should find several frequencies within the bandpass of the ceramic filter where the output voltage at TP_8 peaks, but select the largest peak. Sketch the resulting waveforms.

11. Readjust the frequencies of the intelligence for 20 kHz. Again, adjust the RF carrier frequency so that the detected output signal is a maximum. You should find only two distinct frequencies within the bandpass of the filter where the output voltage at TP_8 is at a maximum. These frequencies are those that cause just the upper sideband or just the lower sideband to fall within the bandpass of the filter.

12. Adjust the audio function generator to produce a 20-kHz triangle wave instead of a sine wave. You should find that the output voltage at TP_8 remains as a sine wave. Also, you should find that the output voltage of the detector at TP_8 will peak at several settings of the RF carrier. This

is because only part of the upper sideband or lower sideband frequencies produced at the balanced modulator output are being passed by the 20-kHz-bandwidth ceramic filter. This should also happen if a square-wave intelligence signal is being used.

13. Now adjust the intelligence signal to produce a 200-Hz triangle wave. Now the first 100 harmonics of the triangle wave's fundamental frequency can be within the 20-kHz bandwidth of either the upper sideband or lower sideband at the output of the balanced modulator. Thus, if we tune the RF carrier so that just the upper sideband or lower sideband frequency component is passed by the ceramic filter, the product detector should re-create the original triangle wave with minimal distortion. Tune the carrier to produce an output signal at TP_8 that is least distorted. Sketch the original and recreated intelligence signals.

14. Repeat step 12 using a square-wave intelligence signal.

15. Repeat step 13 using a square-wave intelligence signal.

QUESTIONS:

1. Draw a sketch of the frequency spectra of the DSB-SC signals produced in steps 7 and 8. Label all important frequencies on the horizontal axis. Explain why a sine wave at 455 kHz was observed at the output of the ceramic filter in each case.

2. Draw a sketch of the frequency spectra of the DSB-SC signals produced in step 12 when a 20-kHz triangle wave or a 20-kHz square wave was used as the intelligence signal. Draw a sketch of the frequency spectra of the output signal of the product detector. Explain why the output signal of the product detector was not a close replica of the input triangle or square-wave intelligence signal in steps 12 and 14.

3. Draw a sketch of the frequency spectra of the DSB-SC signals produced in steps 13 and 15 when a 200-Hz triangle or square wave was used as the intelligence signal. Draw a sketch of the frequency spectra of the output signal of the product detector. Explain why the output signals of the product detector in steps 13 and 15 were close replicas of the input triangle or square-wave intelligence signals.

FM DETECTION AND FREQUENCY SYNTHESIS USING PLLs

OBJECTIVES:

    1. Further familiarization with phase-locked-loop operation.

    2. To be acquainted with two popular applications of phase-locked loops: FM detection and frequency synthesis.

REFERENCE:

Refer to Section 4-5 in the text.

EQUIPMENT:

Dual-trace oscilloscope

Low-voltage power supply (2)

Function generators (2); one must have a VCG input to produce FM

Frequency counter

COMPONENTS:

Integrated circuits: 565 (2), 7493 (2), 7420

Transistor: 2N2222

Capacitors: 0.001 μF (4), 0.1 μF (3), 10 μF

Resistors ($\frac{1}{2}$ watt): 680 Ω (4), 4.7 kΩ (3), 10 kΩ, 33 kΩ (2)

PROCEDURE:

    1. Build the FM detector circuit shown in Figure 11-1. Apply $\pm$ 10 V dc to this circuit and measure the free-running frequency of the VCO part of the 565 PLL by measuring the frequency of the VCO output waveform at TP_4 with a frequency counter.

        free-running frequency = _____

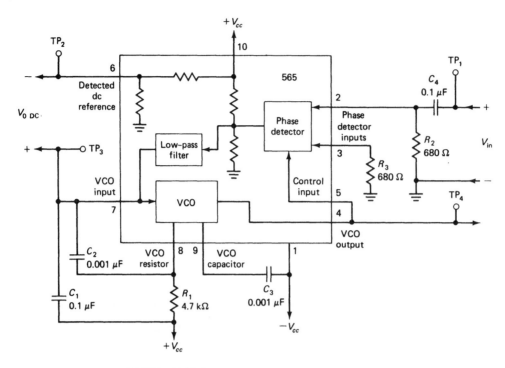

FIGURE 11-1 FM detector using a 565 PLL.

2. Connect a function generator that has VCG capability to the phase detector input at TP_1. This generator will serve as the FM signal generator. Connect channel A of the dual-trace oscilloscope to monitor the output of the FM generator. Connect channel B of the oscilloscope to monitor the VCO output signal at TP_4. Set the FM generator to produce a 500-mV$_{p-p}$ sine wave at the same frequency as the free-running frequency of the VCO part of the PLL. You should see the PLL lock up to the FM generator's frequency by observing both waveforms of the oscilloscope lock up at the same frequency.

3. Connect a second function generator as an intelligence signal by connecting it to the VCG input jack of the RF generator. Set the intelligence signal generator to produce a 100-mV$_{p-p}$ sine wave at a frequency of 20 Hz. You should see both waveforms on the oscilloscope frequency modulate. As long as the deviation of the FM signal does not cause the VCO output of the PLL to exceed its tracking range, the waveforms should remain frequency locked together.

4. Now connect channel A of the scope to monitor the original intelligence signal at the VCG jack of the RF generator. Connect channel B of the oscilloscope to monitor the detected dc reference signal at TP_3. In Figure 11-2, sketch the waveforms observed on the oscilloscope.

5. Increase and decrease the amplitude of the intelligence signal. Note the effect the amplitude changes have on the signals observed at TP_4 and TP_3. Return to the original amplitude setting before proceeding.

6. Increase and decrease the amplitude of the FM signal. Again, note the effect the amplitude changes have on the signals observed at TP_4 and TP_3. Return to the original amplitude setting before proceeding.

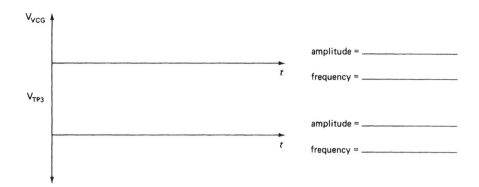

amplitude = _____

frequency = _____

amplitude = _____

frequency = _____

FIGURE 11-2 Waveforms observed on the oscilloscope.

7. Increase the intelligence signal frequency and watch the waveform at TP_3. You will discover a critical frequency where the phase-locked loop can no longer follow the variations in frequency of the RF generator. Record this frequency and return the intelligence generator back to 20 Hz.

8. Set the intelligence generator to produce a 50-mV$_{p-p}$ 20 Hz square wave. Sketch the resulting detected signal at TP_3. It should exhibit overshoot and ringing. Measure the ringing frequency. It should be fairly close to the same frequency as the critical frequency measured in step 7.

9. Build the frequency synthesizer circuit shown in Figure 11-3. Leave jumpers J_1, J_2, J_3, and J_4 disconnected from any test points. Apply ± 5 V dc to this circuit. Again, check the VCO output at TP_9 and measure the free-running frequency.

free-running frequency = _____

10. Connect a function generator to act as a master oscillator at TP_{10}. Connect channel A of the dual-trace oscilloscope at TP_{10} and channel B at TP_9. Connect jumper J_1 to ground. This causes the binary counter to be engaged to count up to its maximum value as a MOD-256 counter. This particular divide-by ratio, N, which is 256 in this case, determines the multiple that the output frequency is with respect to the input frequency. Even though the digital divider is dividing by 256, the value of $N = 256$ causes the input frequency to be multiplied by 256 to create the output frequency. Thus, by providing a constant input frequency through the design of a single, stable oscillator, we can produce a stable but variable output frequency simply by changing the value of N in the digital divider network. This can be done easily through the use of proper digital logic gates and appropriate switches which would be set properly either manually or perhaps by a microcontroller computer. This offers an alternative to the expensive use of multiple-crystal oscillators to create stable but multiple operating frequencies in receiver and transmitter designs. This configuration, shown in block-diagram form in Figure 11-4, is known as indirect frequency synthesis.

Apply a 0.5-V$_{p-p}$ sinusoidal signal to the synthesizer circuit at TP_{10}. Use an input frequency fairly close to the free-running frequency divided by 256. Observe the waveforms of V_{in} and V_o at TP_{10} and TP_9, respectively. Verify that f_o/f_{in} is equal to 256, using a frequency counter to measure f_o and f_{in}. You should notice that the loop locks up for output frequencies within a certain range about the free-running frequency.

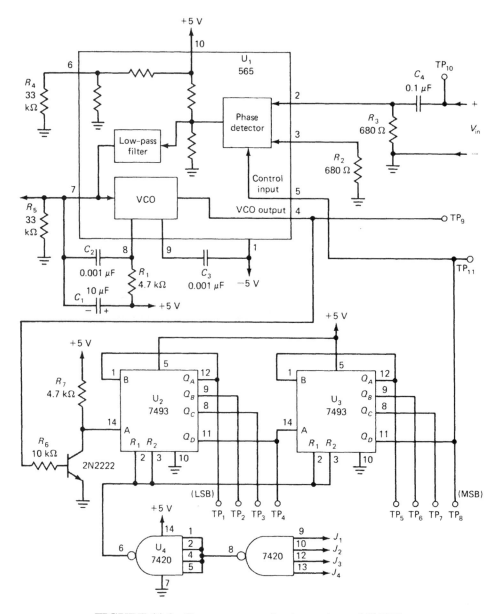

FIGURE 11-3 Frequency synthesizer using a 565 PLL.

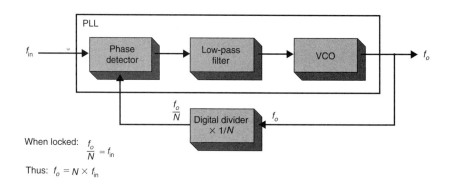

When locked: $\dfrac{f_o}{N} = f_{in}$

Thus: $f_o = N \times f_{in}$

FIGURE 11-4 Frequency synthesizer block diagram.

11. Connect jumpers J_1 to TP_1, J_2 to TP_2, J_3 to TP_7, and J_4 to TP_8. This makes the counter into a MOD-195 counter. If you are unfamiliar with MOD counters and digital dividers, refer to a digital electronics text to review how the divide-by-195 function occurs with the 7420 NAND gate controlling the reset inputs of the dividers. Verify that $f_o = 195 \times f_{in}$ by measuring the frequencies, f_{in} and f_o, with a frequency counter when f_{in} is set at approximately l/195th of the free-running frequency. Again, measure the tracking range of the loop and the duty cycle of the output signal of the digital counter at TP_{11}. Note that it may be possible for the PLL to lock up at input frequency ranges other than 1/195 of f_o. These "false" tracking ranges are usually quite small and unstable in nature.

12. Connect J_1 to TP_1, J_2 to TP_2, J_3 to TP_6, and J_4 to TP_8. Vary the frequency of V_{in} until the phase-locked loop locks up. Again, measure f_o and f_n with the counter and experimentally determine what the value of N now is for the digital divider. Determine theoretically what N is by following one of the procedures:

   a. Review what signals cause the NAND gate to reset the counter, express this as a digital count in binary notation, and convert to decimal notation.

   b. If you are patient and determined, you may be able to carefully count how many pulses of the input signal to the digital counter it takes to produce one complete pulse at the output. If your oscilloscope has delayed sweep capability, it should be helpful to you in your counting process!

Again, measure the duty cycle of the output signal of the digital divider and the tracking range of the synthesizer. Be prepared to demonstrate to the lab instructor that you understand how the frequency synthesizer operates.

13. Try a small-value duty-cycle case. Connect the jumpers properly for a divide-by-139 counter. Calculate the free-running frequency divided by 139 and set f_{in} at this frequency value. Slowly vary the frequency, f_{in}, until you notice that the loop is back in lock. Again, measure f_o and f_{in} to verify that $N = 139$. Measure and record the duty cycle and tracking range as done in previous steps. You should find that at low duty cycles, the tracking range is very narrow, making the system more unstable.

QUESTIONS:

1. Describe how the phase-locked loop can be used successfully to detect an FM signal. Explain how it works.

2. Explain why in step 6 there was very little if any effect of changing the amplitude of the FM signal on the amplitude of the detected output signal.

3. Describe how the phase-locked loop can be used successfully to synthesize a range of output frequencies by using a single stable oscillator and digital counter. Explain how the circuit works. Also, state its advantage over the use of multiple-crystal oscillators in a "channelized" radio.

4. Research: A limitation of this synthesizer design is its noncontinuous frequency ranges that can actually be synthesized. How can this limitation be removed through the addition of mixers? (Refer to Section 4-5 in your text.)

NAME _____

GENERATING FM FROM A VCO

OBJECTIVE:

To use a voltage-controlled oscillator (VCO) to develop a frequency-modulated (FM) signal.

REFERENCE:

Refer to Sections 3-3 and 4-5 of the text.

EQUIPMENT:

Function generator

Spectrum analyzer, with a resolution bandwidth of 3 kHz or less

dc power supplies (2)

COMPONENTS:

Mini-Circuits®, Model ZX95-100, voltage-controlled oscillator

Mini-Circuits®, Model SM-BF50, adapter, SMA-M to BNC-F

50 Ω coaxial cable, 3 foot, BNC-M connectors (2)

Banana connector jumper cables, 3 foot (6)

INTRODUCTION:

A ZX95-100 is a voltage-controlled oscillator that can be tuned from 50 MHz to 100 MHz by changing the voltage at the Vtune port from 0.5 V to 17 V. Another dc voltage of +12 V must be applied to the Vcc. The modulating signal is provided via a 10 kΩ (10 turn) potentiometer and a 10 μF capacitor.

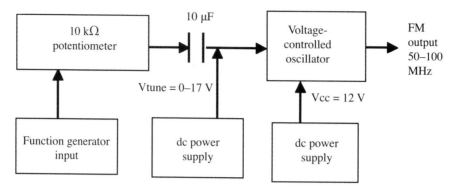

FIGURE 12-1 VCO assembly.

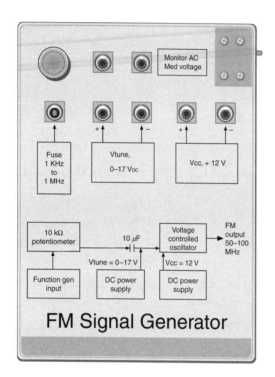

FIGURE 12-2 FM signal generator assembly, front view.

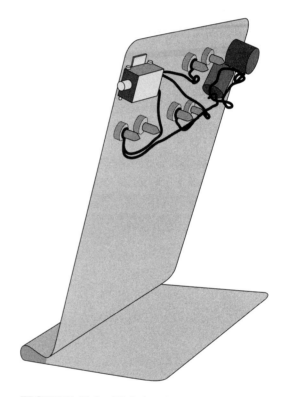

FIGURE 12-3 FM signal generator assembly, side/back view.

PROCEDURE:

1. Connect the RF OUT port of the VCO to the spectrum analyzer via a coaxial cable with BNC connectors.

2. Set the spectrum analyzer center frequency to 50 MHz and set the span at 10 MHz/div.

3. Apply +12 V dc to the Vcc port of the VCO.

4. Observe the signal on the spectrum analyzer at 50 MHz.

5. Connect another dc power supply to the Vtune port of the VCO. Observe the spectrum analyzer display while increasing the dc voltage until the center frequency is 70 MHz.

6. Adjust the spectrum analyzer center frequency so that the signal is in the middle of the screen.

7. Connect the function generator to the Vtune port via a 10 kΩ potentiometer and a 10 μF capacitor, as shown in Figure 12-1.

8. Adjust the frequency of the function generator to 100 kHz and the amplitude to the maximum value.

9. Observe the spectrum analyzer display and start reducing the span until you reach 100 kHz/div.

10. Adjust the amplitude of the function generator from a very low value to a very high value, and note that the frequency spectrum varies as illustrated in Figure 3-4 of the textbook. Further note that the amplitude value of the carrier and each sideband follow the Bessel Functions shown in Table 3-1 in the textbook.

11. Measure and record the rms voltage of the 100 kHz signal at the first J_0 null ($m_f = 2.4$). _____

12. Measure and record the rms voltage of the 100 kHz signal at the 2nd J_0 null ($m_f = 5.5$). _____

13. Record the ac voltage of the 100 kHz modulating input voltage when $m_f = 5.5$. _____

14. Calculate the ac voltage for $m_f = 5.0$. _____

15. Set the ac voltage at the level calculated in step 14.

16. Look at section 3-3, Table 3-1 of the textbook and record the J_0–J_8 in Table 12-1.

17. Using example 3-5 from the text as a guide, convert J_0–J_8 to dB and record in Table 12-1.

18. With m_f set at 5 for the VCO, measure and record the values of J_0–J_8 in Table 12-1.

Note that J_0–J_5 can be measured directly off the spectrum analyzer. To measure J_6–J_8, the center frequency can be changed from 70.01 MHz to 70.31 MHz.

TABLE 12-1   Calculations and Measurements for VCO Project. (Bessel Functions for $x = 5$)

	J #	VALUE FROM TABLE 3-1 IN THE TEXTBOOK	CALCULATED (dB)	MEASURED (dBm)	MEASURED MINUS CALCULATED VALUES
1	J_0		0 ref	Ref:	_____
2	J_1				
3	J_2				
4	J_3				
5	J_4				
6	J_5				
7	J_6				
8	J_7				
9	J_8				

QUESTIONS:

1. Explain the process by which a dc voltage causes the frequency of a voltage-controlled oscillator (VCO) to shift. Which component within the VCO is controlled to affect VCO frequency?

2. What happened to the carrier in steps 11 and 12? Compare and contrast the distribution of power among carrier and sidebands in FM versus AM transmissions.

3. Describe a practical application of the behavior depicted in steps 11 and 12. Specifically, describe how the Bessel null condition can be used to verify modulator linearity.

4. What is the purpose of the 10 μF capacitor in this setup?

5. Research: what are the similarities and differences between direct and indirect FM? Which type of modulation is embodied in the setup depicted in this experiment?

6. How would this experiment need to be modified to represent the form of angle modulation that is the contrast to the form identified in question 5?

PULSE AMPLITUDE MODULATION

OBJECTIVES:

1. To become familiar with pulse-amplitude modulation (PAM) techniques.

2. To test and evaluate a simple PAM modulator and demodulator.

3. To observe the effects of aliasing distortion on sampled signals.

REFERENCE:

Refer to Sections 7-2 and 7-3 in the text

EQUIPMENT:

Dual-trace oscilloscope

Low-voltage power supply

Function generator

Multimeter

Breadboard

COMPONENTS:

555 timer

10-kΩ potentiometer

2.2-nF capacitor, 10-nF capacitor (2 each), 10-μF capacitors

330-Ω, 1-kΩ, 10-kΩ resistors

4001 diode or equivalent

2N2222 NPN transistor or equivalent

INTRODUCTION:

This experiment introduces the principles of pulse modulation and demodulation and demonstrates how undesired alias signals are produced.

Pulse-amplitude modulation (PAM) is one form of pulse modulation used in the transmission of digital signals in a message-processing format. In PAM, the RF carrier pulse's amplitude is directly proportional to the data signal's amplitude (Figure 13-1).

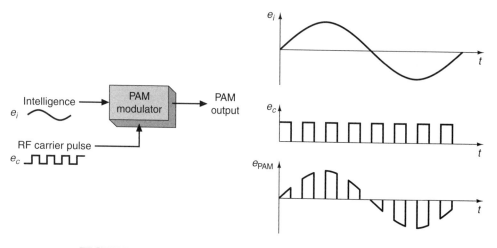

FIGURE 13-1   Generation of pulse-amplitude modulation.

PAM is the least desirable of the various types of pulse modulation used in digital communication because amplitude variations are easily subject to noise interference, which is also an amplitude-varying phenomenon. Other pulse-modulation schemes, to be studied in subsequent experiments, avoid noise interference by relying on frequency or phase shifts to represent information, or by representing sampled amplitudes as multiple-bit digital words.

The simplest form of the PAM modulator is an analog switch that is turned on and off at the RF carrier pulse rate. As this switch changes state, the intelligence (data) signal is connected and disconnected from the output. Thus the output PAM signal is a sampled version of the input intelligence signal. A spectral analysis of the complex PAM signal reveals that the data-signal frequency components are far removed in frequency from the other frequency components of the complex PAM waveform.

This is true only if the sampling rate is considerably higher than the highest data frequency being used. Thus if the sample rate is kept high enough in comparison to the data signals being used, a simple low-pass filter can be used as a PAM demodulator to retrieve the original information.

PROCEDURE:

Part I: PAM Modulator

Using the datasheet for the 555 timer operating as an ASTABLE oscillator, determine the possible range of output frequencies for the square-wave signal produced by the oscillator in Figure 13-2.

1. Build the 555 astable oscillator shown in Figure 13-2.
2. Apply 5 V and observe the output pulse on the oscilloscope. Adjust the 10-kΩ potentiometer until there is a 50% duty cycle and a 30–35 kHz output signal. This signal will serve as the RF carrier. Measure its frequency and amplitude.

Frequency _____

Amplitude _____

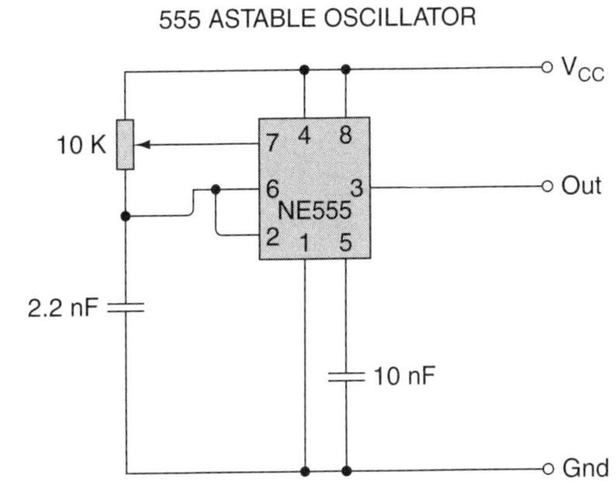

555 ASTABLE OSCILLATOR

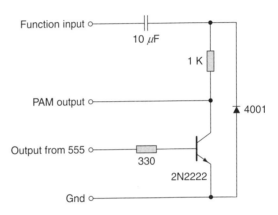

FIGURE 13-2 555 timer operating as an astable oscillator.

3. Add the PAM modulator circuit to the output of the 555 oscillator as shown in Figure 13-3 below.

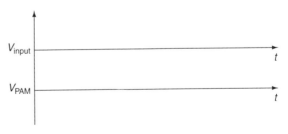

FIGURE 13-3 PAM modulator.

4. Apply a 3 kHz sine wave to the function input. This sine wave will function as the data signal. Observe the signal at the function input on Channel A of your oscilloscope and the resulting PAM output on Channel B. Let the data signal serve as the trigger for the oscilloscope. Sketch the signal of your PAM output and your function input on the graph shown in Figure 13-4.

V_{input} t

V_{PAM} t

FIGURE 13-4 Observed waveforms at function input and PAM output.

Determine the frequency and amplitude of the two signals and record the data below. In the case of the PAM signal, estimate the frequency of the overall envelope of the signal.

Input Frequency _____   PAM Frequency _____

Input Amplitude _____   PAM Amplitude _____

Check that your scope is dc coupled and measure the dc offset between the two signals.

dc Offset _____

5. Apply a 20-kHz sine wave to the function input. Repeat the measurements you made above. You should notice considerable distortion resulting from the signal frequncy being too close to the sampling rate. This is known as *aliasing distortion*. Sketch the signal of your PAM output and your function input on the graph shown in Figure 13-5 below.

FIGURE 13-5   Observed input and output signals with aliasing distortion.

6. Determine the frequency and amplitude of the input and PAM Signal at ~20 kHz.

Input Frequency _____   PAM Frequency _____

Input Amplitude _____   PAM Amplitude _____

dc Offset _____

Part II: PAM Demodulation

7. To recover the PAM signal as an analog signal, we will pass it through a low pass filter. Add the low pass filter circuit to your previous circuit as shown in Figure 13-6.

8. Apply 3 kHz and 20 kHz sine waves to the function input. Sketch the function input and your PAM demodulation output on the graphs in Figures 13-7 and 13-8.

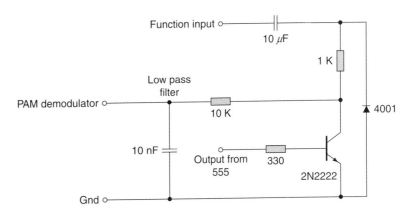

FIGURE 13-6 PAM demodulator.

FIGURE 13-7 Graph for function input and PAM demodulation output at 3 kHz.

FIGURE 13-8 Graph for function input and PAM demodulation output at 20 kHz. Note: You should observe substantial aliasing in this signal.

9. Determine what would be considered the highest data frequency that can be used with this PAM circuit. There are several metrics for this, but you might observe a Nyquist frequency (half of your original carrier rate) and see how this compares to a low frequency. A commonly used definition is that a signal is considered distorted when the noise amplitude exceeds 10% of the total peak-to-peak amplitude of the waveform.

$f_{max} =$ _____

10. As time permits, double the carrier frequency of your oscillator and repeat your measurements in steps 1–9.

QUESTIONS:

1. How does the input frequency compare to the PAM frequency? Should these values be the same? Why or why not?

2. How does the amplitude of the input signal compare to the PAM amplitude? Should these values be the same? Why or why not?

3. Was there a dc offset between the PAM signal and the input signal? Explain where this would come from.

4. What is the cutoff frequency of the low pass filter used? What impact does this have on the signal?

5. Why is the observed voltage of the demodulated signal lower than the input signal?

6. What impact does aliasing have on the demodulation of the signal?

7. What is the highest data frequency possible for this circuit? How does this compare to the Nyquist frequency? What are the limits of the Nyquist criteria?

8. What would be the impact of doubling the carrier frequency?

TIME-DIVISION MULTIPLEXING

OBJECTIVES:

    1. To become familiar with the use of time-division multiplexing (TDM) techniques.

    2. To observe pulse width modulated (PWM) signals.

    3. To test a TDM communication system that uses PWM signals.

REFERENCE:

    Refer to Section 7-4 in the text.

EQUIPMENT:

    Dual-trace oscilloscope

    Function generator

    Digital multimeter (DMM)

    Breadboard

COMPONENTS:

    Arduino microcontroller

    10-kΩ potentiometers

    10-kΩ resistor

    Push-button switch suitable for breadboard

INTRODUCTION:

    Pulse-width modulation (PWM) is often used to transmit low-frequency signals over long distances through telephone lines or fiber-optic cables. In PWM the amplitude of the analog signal is converted to variations in pulse length of the carrier signal as shown in the middle row of Figure 14-1.

    With pulse-modulation schemes the time intervals between pulses can be filled with samples of other messages. In other words, a form of channel sharing known as *time-division multiplexing* (TDM) can be deployed. In general terms, multiplexing involves conveying two or more information signals over a single transmission

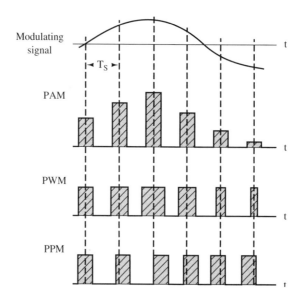

FIGURE 14-1 Types of pulse modulation.

channel. In TDM, each information signal accesses the entire channel bandwidth but for only a small part of the available time. TDM is analogous to computer time-sharing, where several users make use of a computer simultaneously. TDM is extensively used in both wired and wireless communication systems, especially telephone networks.

In the first part of this experiment the representation of an analog signal as a PWM signal is demonstrated. A common application of these PWM signals is in remote control vehicles where the servos and speed controllers respond to PWM signals.

In part two of the experiment, a communications protocol is developed in which a set of PWM signals is time-division multiplexed into a single line. This allows multiple signals to be carried on the same cable.

PROCEDURE:

Part I: Pulse Width Modulation

1. You will be using the Arduino microcontroller. A microcontroller is a small computer with programmable input and output. In this experiment you will use a preprogrammed Arduino that acts as a PWM generator. Connect the Arduino to the elements in your breadboard as shown in Figure 14-2. Be sure to connect the +5 V and the ground from the Arduino to your breadboard.

2. If your Arduino is not preprogrammed, open the Arduino IDE on your computer and load the program "ArduinoTDM." (See the Appendix for the program code.) Press the "Compile and download" button to program your microcontroller.

3. Connect your oscilloscope channel B to D4 on the Arduino. Adjust the potentiometer from one extreme to the other and observe the changes on the oscilloscope. Measure the maximum and minimum pulse widths and

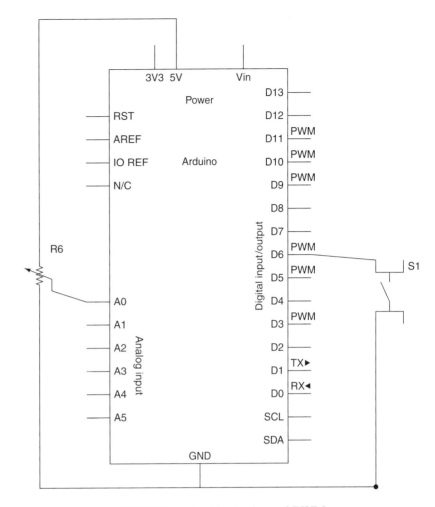

FIGURE 14-2 Single channel PWM.

calculate the range of the PWM signal by taking the difference between the maximum and the minimum.

Maximum Pulse Width = _____ **μs**

Minimum Pulse Width = _____ **μs**

Range = _____ **μs**

4. Measure the range of input voltages at pin A0 with your DMM. Calculate the conversion from pulse width to voltage by taking the ratio of the PWM range over the voltage range.

   μs/V = _____

5. Now the dc input voltage to the ADC will be replaced with an analog signal. Disconnect the wire to the potentiometer and replace it with the function generator signal. Adjust the function generator to produce a 2-V peak-to-peak sine wave centered at 1 V at a very low frequency of 2 Hz. You should see the pulse width of the PWM signal change as the wave's instantaneous amplitude varies.

6. Increase the frequency of the input signal to about 60 Hz and sketch the resulting signal in Figure 14-3. Connect your oscilloscope probe A to the function generator so you can compare the input signal to the output signal.

FIGURE 14-3   Observed input signal at 60 Hz and PWM output.

7. Determine the time between pulse frames for the PWM signal. This time is the same as the sampling rate of the PWM encoder and provides a limitation on the PWM scheme.

Time = _____ μs

8. Increase the frequency of the input signal to 500 Hz and sketch the resulting signal in Figure 14-4. At 500 Hz, the time between cycles is 2000 μs.

FIGURE 14-4   Observed PWM output signal with 500-Hz input.

Part II: Time-Division Multiplexing (TDM)

9. You will be using the same Arduino microcontroller and program as in the previous part. In the previous experiment we looked at converting a single analog voltage to a PWM signal. Now a series of six analog signals will be time-division multiplexed onto a single line. Build the circuit shown in Figure 14-5 on your breadboard.

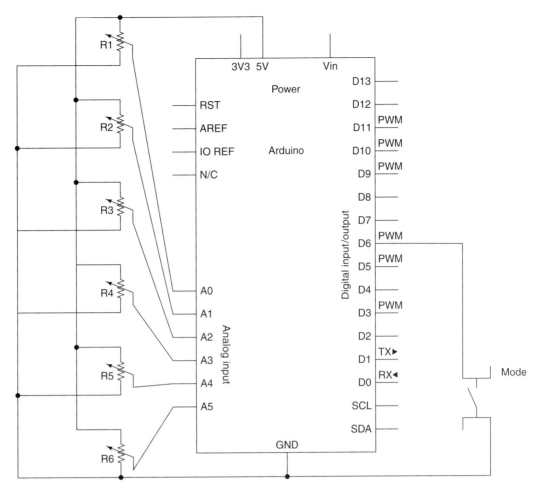

FIGURE 14-5 TDM demonstration circuit.

10. Connect channel B of your oscilloscope to D4 on the Arduino. Turn each of the potentiometers to the center position. Press and hold the PWM Mode button to switch to TDM mode. Observe and sketch the resulting signal in Figure 14-6.

FIGURE 14-6 Observed input and TDM waveforms.

11. Adjust each of the potentiometers to the value indicated in the table that follows and use the oscilloscope and your multimeter to complete Table 14-1.

TABLE 14-1   Observed Pulse Widths as a Function of Input Voltage.

POTENTIOMETER #	VOLTAGE (V)	MEASURED PULSE WIDTH (μs)
1	1.5	
2	2.9	
3	.8	
4		950
5		800
6		62

12. With the potentiometers in the positions given in the table above, sketch the resulting TDM waveform.

13. Now two of the dc input voltages from the potentiometers will be replaced with analog signals. Disconnect the wire to the potentiometer A0 and replace it with a 2-V peak-to-peak sine wave centered at 1 V at 60 Hz. Disconnect the wire to the potentiometer A3 and replace it with a 2-V peak-to-peak triangle wave centered at 1 V at 40 Hz. Observe and sketch the resulting signal in the space below.

QUESTIONS:

1. What are the main limitations of the PWM system used in this experiment? What are the limitations on the frequency of the analog signal that can be reliably encoded?

2. What could you change to increase the frequency of the analog signal that can be encoded as a PWM signal?

3. Provide a brief explanation of how TDM can be used to send more than one channel of information over one cable.

4. What are the main limitations of the TDM system investigated in this experiment?

5. Explain the tradeoffs in encoding more signals on a TDM system with respect to the maximum frequency of the analog signal that can be encoded.

NAME _____

PULSE-WIDTH MODULATION AND DETECTION

OBJECTIVES:

> **1.** To observe two types of pulse-time modulation (PTM) waveforms.
> **2.** To test and evaluate a pulse-width modulator.
> **3.** To test and evaluate a pulse-width demodulator.

REFERENCE:

> Refer to Section 7-2 in the text.

EQUIPMENT:

> Dual-trace oscilloscope
> Low-voltage power supply (2)
> Function generator (2)
> Volt-ohmmeter
> Frequency counter

COMPONENTS:

> Integrated circuits: 565, 1496P, 7486
> Transistor: 2N2222
> Capacitors: 4.7 nF (2), 0.01 μF (2), 0.1 μF (6), 0.47 μF (2), 10 μF, 470 μF
> Resistors ($\frac{1}{2}$ watt): 47 Ω, 100 Ω, 390 Ω (2), 680 Ω, 1 kΩ (5), 3.3 kΩ, 4.7 kΩ (2), 5.6 kΩ, 10 kΩ (4), 33 kΩ, 47 kΩ
> Potentiometers: (10-turn trim) 5 kΩ (2)

INTRODUCTION:

> A type of modulation often used when transmitting low-frequency signals over a long distance via telephone lines or fiber optic scale is pulse-time modulation (PTM). In PTM, the amplitude of the intelligence signal is converted to variations in pulse length or variations in pulse position of the carrier signal. These two types of PTM are referred

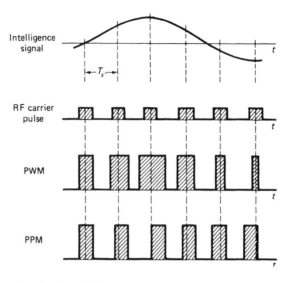

FIGURE 15-1 PWM and PPM waveforms.

to as pulse-width modulation (PWM) and pulse-position modulation (PPM), respectively. The wave shapes of PWM and PPM are given in Figure 15-1.

The amplitude of the carrier pulses remains constant in either PWM or PPM. Thus PTM signals exhibit the same advantages that are exhibited by FM signals: noise immunity and low distortion. In addition, pulse signals are easily reproduced if they do get noisy and distorted. This is not possible if analog signals are used.

One possible method of generating PWM and PPM uses the phase-locked loop and digital exclusive-or gate illustrated in Figure 15-2. The PLL is designed to lock up near the frequency of the carrier. The carrier frequency is applied to the phase detector input, thus causing the detector to lock up at the carrier frequency. The intelligence signal is applied to the reference input of the VCO and causes the VCO's output signal to shift its phase with respect to the reference phase by an amount proportional to the amplitude of the intelligence signal. This fits the definition of pulse-position modulation (PPM). If the VCO output signal is fed into a switching transistor, the output of the switching transistor can be described as being a TTL-compatible PPM signal. To create PWM, the original RF carrier pulse and the PPM signal are fed into the two inputs of an exclusive-or gate. Its output will be "high" only during the time when the RF carrier pulse and PPM signals are at different logic states. Since the two signals are synchronized by the PLL, the exclusive-or output signal will be "high" only during the time of phase difference between the RF carrier pulse and the PPM pulse. This creates PWM. Refer to Figure 15-1 to verify this.

A method that successfully demodulates the PWM signal uses a 1496 balanced modulator as given in Figure 15-4. If the 1496 product detector that was investigated in Experiment 10 is slightly altered so that the differential amplifier transistors act as switching transistors, the pulse signal produced at its output will exhibit a dc offset that will vary as a function of the phase difference between its two inputs. Thus if the two input signals are the PWM signal and the original RF carrier pulse, the output voltage of the product detector will resemble the original intelligence signal. Fortunately, all of the other frequency components produced by the mixing action of the switching transistors in the 1496 are much higher frequencies than the original intelligence signal. Therefore, if the complex output signal of the 1496 is passed through a simple low-pass filter, the original intelligence signal will be recreated.

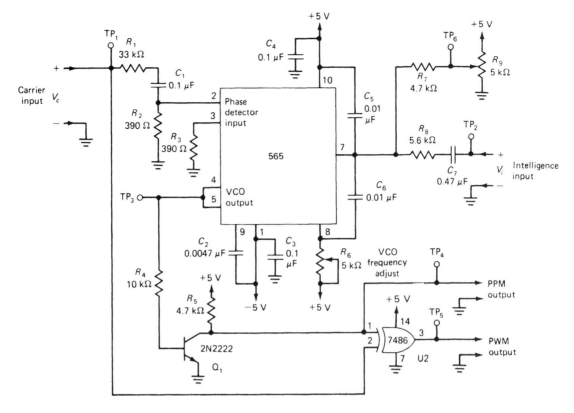

FIGURE 15-2 PWM modulator stage.

PROCEDURE:

1. Calculate the free-running frequency of the PLL found in the PWM modulator of Figure 15-2 if R_6 is adjusted to 3 kΩ. R_6 is the timing resistor, and C_2, found on pin 9, is the timing capacitor. The free-running frequency, f_0, is approximated with the following relation, as determined from the manufacturer data sheet:

$$f_0 \cong \frac{1}{3.7R_6C_2}$$

2. Build the circuit given in Figure 15-2. Apply ±5 V dc and measure the VCO output frequency at TP$_3$. Adjust R_9 so as to produce approximately 3.5 V dc at TP$_6$. Adjust R_6 to produce exactly 15 kHz at TP$_3$.

3. Apply a 3 V, 15 kHz positive square wave at TP$_1$. This will function as the RF carrier pulse that is to be modulated by this circuit. Observe the waveform at TP$_3$. Use 10:1 scope probes to avoid loading of the signal by the scope. You should be able to see the PLL lock up as the RF carrier frequency is adjusted near 15 kHz.

4. Set the RF carrier back at 15 kHz. This should be approximately in the middle of the tracking range. If not, readjust R_6 and repeat step 3 to make it so. Now apply a 2-V$_{p-p}$, 2-kHz sine wave at TP$_2$. This will function as the intelligence signal. Observe the signal at TP$_3$. Notice that as the intelligence signal amplitude is varied, the display will blur. This is due to the

intelligence signal introducing phase shift into the PLL's VCO's output signal. This is PPM.

5. Observe the waveform at TP$_4$ with channel A of the oscilloscope. This is a TTL-compatible PPM waveform. Now observe the intelligence signal at TP$_2$ with channel A and observe the PPM signal at TP$_4$ with channel B. Trigger the oscilloscope with the intelligence signal. Increase the amplitude of the intelligence signal to approximately 5 V$_{p-p}$. Carefully adjust the trigger level of the oscilloscope and the frequency of the intelligence signal to produce a stable display. Sketch the resulting display in Figure 15-3. It should look similar to the sketch given in Figure 15-1.

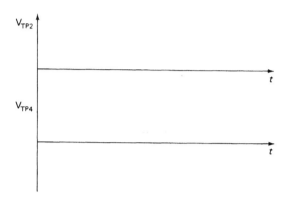

FIGURE 15-3   Intelligence signal and TTL-compatible PPM waveform.

6. Temporarily disconnect the intelligence signal from TP$_2$. Observe the output of the exclusive-or gate at TP$_5$ with channel B and observe the original RF carrier pulse at TP$_1$ with channel A. Let the RF carrier pulse serve as the trigger signal for the scope. The waveform at TP$_5$ should be a square wave at exactly twice the frequency of the RF carrier pulse. Fine-tune the frequency of the RF carrier pulse. It should vary the duty cycle of the output waveform at TP$_5$. Adjust the frequency such that a duty cycle of approximately 50% is produced. Measure the frequency of the RF carrier pulse. It should be approximately in the center of the PLL's tracking range.

7. Reconnect the intelligence signal at TP$_2$. Adjust the amplitude between zero and 5 V$_{p-p}$. You should notice that the intelligence signal causes the pulse duration of the output signal of the exclusive-or gate to vary. This is seen as a blurring effect on the waveform at TP$_5$. This is PWM.

8. Move the channel A probe to TP$_2$. Observe the waveform at TP$_5$ with channel B with the intelligence signal serving as the trigger signal for the oscilloscope. Adjust the amplitude of the intelligence signal to approximately 2 V$_{p-p}$. As in step 5, carefully adjust the trigger level of the scope and the frequency of the intelligence signal in order to produce a stable display. Sketch the resulting PWM display. It should look similar to the sketch given in Figure 15-1. Do not disassemble this circuit before proceeding to step 9.

9. Build the PWM demodulator circuit given in Figure 15-4. Connect the output signal of the PWM modulator to the input of the PWM demodulator by connecting a jumper between TP$_5$ and TP$_8$. Also, provide the balanced modulator with the RF carrier pulse input by connecting a jumper between

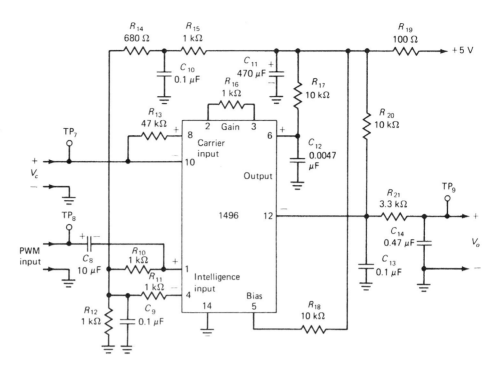

FIGURE 15-4 PWM demodulator.

TP$_1$ and TP$_7$. Apply ±5V dc to both circuits. The RF carrier should still be set at 3-V amplitude at a frequency near 15 kHz that causes approximately 50% duty cycle to exist at the modulator output at TP$_5$. The balanced modulator may capacitively load the RF carrier pulse, so do not expect the waveform at TP$_1$ to remain as a clean square wave as was observed in preceding steps.

10. Adjust the amplitude of the intelligence signal to 1 V$_{p-p}$ at a frequency of 30 Hz. Observe the waveform at TP$_9$. It should be a clean replica of the original intelligence signal. Fine-tune the amplitude and frequency of the carrier signal to produce optimum results. Verify that the system is working properly by temporarily disconnecting each of the two jumpers connected in step 9, one at a time. The output waveform of the demodulator at TP$_9$ should disappear if either of the two jumpers is disconnected. With the jumpers reconnected, adjust the frequency of the RF carrier pulse above and below the PLL's tracking range as was measured in step 3. You should see that when the PLL goes out of lock, the output signal of the demodulator at TP$_9$ disappears again. Readjust the frequency of the RF carrier pulse such as to produce 50% duty cycle at TP$_5$ before proceeding.

11. Determine the range of the intelligence frequencies that can be reproduced successfully by the PWM demodulator without any resulting distortion. Take data so as to be able to sketch the frequency response of the PWM digital communication system, that is, V_{TP8} vs. frequency.

12. Determine the maximum and minimum amplitudes of both the intelligence signal and the RF carrier pulse that allow the output signal at TP$_9$ to remain undistorted.

QUESTIONS:

1. Provide a brief explanation of how the intelligence signal applied at TP_2 of Figure 15-2 is encoded as variations in pulse width in the output waveform at TP_5.

2. How does the 1496 balanced modulator stage shown in Figure 15-4 function to detect the original information signal from the encoded PWM signal applied at TP_8?

3. What are the main limitations of the PWM communication system investigated in this experiment?

INTRODUCTION TO ANALOG-TO-DIGITAL CONVERSION (ADC) AND DIGITAL-TO-ANALOG CONVERSION (DAC)

OBJECTIVES:

1. To build and test a working analog-to-digital converter (ADC).
2. To build and test a working digital-to-analog converter (DAC).
3. To explore the effects of sample rate and resolution on the ability of an ADC and DAC to reproduce signals.

REFERENCE:

Refer to Section 7-4 in the text.

EQUIPMENT:

Dual-trace oscilloscope
Low-voltage power supply
Function Generator
Digital Multimeter
Breadboard

COMPONENTS:

Arduino microcontroller
Multi-turn 10-kΩ potentiometer
100-nF capacitor
Resistors: 330 Ω (8 each), 1.5-kΩ (8 each), 3-kΩ (8 each), 10-kΩ (2 each)
LEDs (8 each)
Pushbutton switch

INTRODUCTION:

The first stage in any digital communication system is to convert the analog data into a digital word where the numerical value of the word is proportional to the analog input. This is done with an analog-to-digital converter (ADC). ADCs are found either as standalone integrated circuits or as part of microcontrollers. Once the ADC has sampled the analog signal into a digital format, the data are coded, or prepared for transmission over a communications channel. After transmission, data are converted back into an analog format via a digital-to-analog converter (DAC). The DAC produces a voltage that is proportional to the digital word applied to its input; in other words, it performs the inverse function of the ADC.

The first part of this experiment develops an elementary coding scheme using an 8-bit ADC. The specific code being used is to let the digital word's binary value represent the amplitude of the applied analog signal when the sample is taken by the ADC. With 8-bit ADCs there are 2^8, or 256, possible binary values and specific analog levels. The eight values will be displayed on a series of light emitting diodes as shown in Figure 16-1.

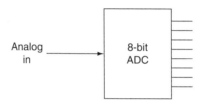

FIGURE 16-1 8-bit ADC.

In the second part of the experiment, the analog output voltage level is reconstructed using a digital-to-analog converter (DAC) in which the voltage output is based on the binary value of the applied data. When the two systems are combined, the original signal can be reconstructed. The data can be transported over long distances between the ADC and DAC as digital information as illustrated by Figure 16-2.

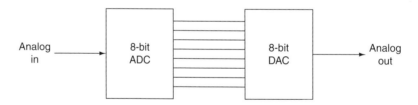

FIGURE 16-2 Data transport between ADC and DAC systems.

PROCEDURE:

Part I: Analog to Digital Conversion (ADC) and Simple Pulse-Code Modulation (PCM)

1. Determine the cutoff frequency of the low pass filter used in Part II, step 4.
2. You will be using the Arduino microcontroller. A microcontroller is a small computer with programmable input and output. In this case you will use a program to have the Arduino act as an 8-bit parallel ADC. Build the circuit shown in Figure 16-3 below. Be sure to connect the +5 V and the ground connections from the Arduino to your breadboard.

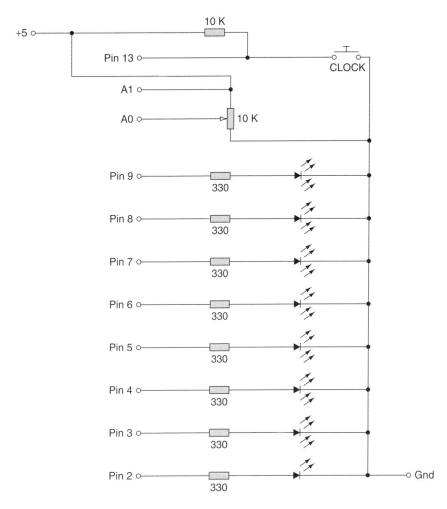

FIGURE 16-3 8-bit parallel ADC.

3. If your Arduino is not preprogrammed, open up the Arduino IDE on your computer and load the program "ArduinoParallelADC." Press the compile and download button to program your microcontroller.

4. Press the CLOCK button. When the button is pressed, it updates the LEDs at 100 Hz. Slowly increase the analog voltage to pin A0 from 0 V to approximately 5 V dc by turning the multi-turn 10-kΩ potentiometer while holding down the CLOCK button. You should see the eight LEDs count from 00000000 to 11111111. This is a very simple pulse-code modulation (PCM) coding scheme, as the binary code of the parallel output digital data of the ADC represents the amplitude of the analog input voltage. Determine the dc level for V_{in} to produce a digital count of 010000000 by using your multimeter to measure the voltage between A0 and ground. Repeat for digital counts of 000000000 and 10101001.

V_{in} **(00000000)** = _____ **V dc**

V_{in} **(01000000)** = _____ **V dc**

V_{in} **(10101001)** = _____ **V dc**

5. Determine the resolution of the ADC. Resolution is the amount of dc variation that causes the smallest amount of change in the digital output code. Measure this by looking at some large changes and taking the ratio of the number of ADC counts to voltage change.

resolution = _____ V dc

6. Now the dc input voltage to the ADC will be replaced with an analog signal. Disconnect the wire to the potentiometer and replace it with the function generator signal. Set the generator's offset control to produce a dc voltage level equal to the value measured in step 2 that produced a count of 1000000 (mid-range). Adjust the function generator to produce a 2-V peak-to-peak sine wave centered at 1 V at a very low frequency of 0.1 Hz. You should see the count of the eight light-emitting diodes (LEDs) change as the sine wave's instantaneous amplitude varies. Slowly increase the amplitude of the sine wave until it causes the LEDs to count from a minimum of 0000000 to a maximum of 10111111, to represent the negative

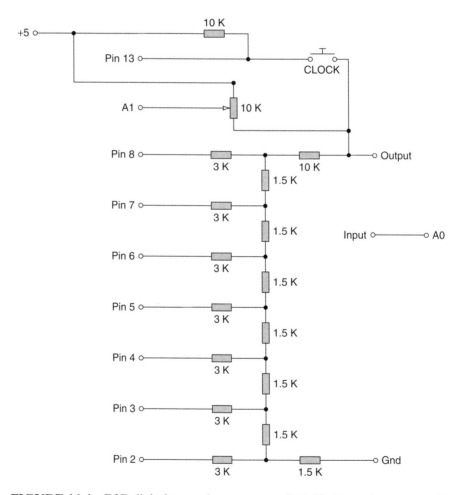

FIGURE 16-4 R2R digital-to-analog converter (DAC). Note: Any reasonably high values for resistance can be used as long as the R/2R relation is preserved.

and positive peaks of the sine wave, respectively. Measure the amplitude of the voltage (V_{in}) that causes this to happen.

$$V_{in} \text{ (max)} = \underline{\hspace{2cm}} V_{p\text{-}p} \text{ dc}$$

Part II: Digital-to-Analog Conversion (DAC)

7. You will be using the same Arduino microcontroller and program as in the previous part. The built-in ADC on the Arduino will be interfaced to an R2R DAC. On your breadboard, build the circuit shown in Figure 16-4. Be certain to connect the +5V and the ground from the Arduino to your breadboard.

8. Adjust the potentiometer to the central position. Apply a 20-Hz , 2-V p-p signal to A0. Observe the signal at the function input on channel A of your oscilloscope and the resulting DAC output on channel B. Let the input signal serve as the trigger for the oscilloscope. Sketch the signal of your DAC output and your function input on the graph shown in Figure 16-5.

FIGURE 16-5 20-Hz input signal and DAC output.

9. The potentiometer connected to A1 is acting to set the sample rate of the signal. When the potentiometer is in one position, the sample rate is set to ~100 Hz; in the opposite position it is set to ~2 kHz. Observe and sketch the 2-V p-p signal from the function generator at a sampling rate of ~100 Hz and ~2 kHz in the space provided in Figure 16-6.

FIGURE 16-6 Observed DAC outputs at 100 Hz and 2 kHz.

10. The digital steps of the waveform can be filtered out with a low-pass filter made by adding the 100-nF capacitor between the output and ground as shown in Figure 16-7.

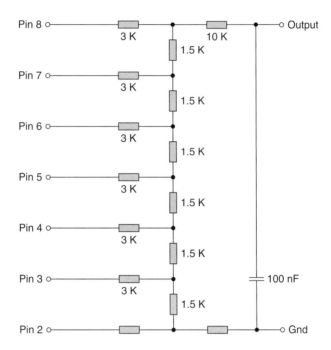

FIGURE 16-7 Filter capacitor added to DAC output.

11. Set the sample rate to ~1 kHz and the input to be a 20-Hz, 2-V p-p signal centered at 1 V. Move channel 2 of the oscilloscope to the output of the low pass filter and sketch the input waveform and the filtered waveform in Figure 16-8.

FIGURE 16-8 Input and filtered waveforms at 1-kHz sample rate.

12. Adjust your sample rate to go as high as possible. (Note that if it is too high, it will be faster than the Arduino can sample and you will obtain noise on the output.) The filtered DAC output should be a close match to the frequency generator input. Observe what happens when V_{in} is changed to a 20-Hz triangle wave and a 20-Hz square wave and sketch the results in Figure 16-9.

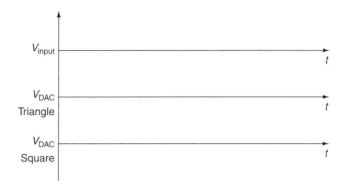

V_{input}　　　　　　　　　　　　　　　　　　　　t

V_{DAC}
Triangle　　　　　　　　　　　　　　　　　　　t

V_{DAC}
Square　　　　　　　　　　　　　　　　　　　　t

FIGURE 16-9　Input and filtered waveforms for 20-Hz
triangle and square waves.

13. Switch back to a sine wave and increase the frequency of the signal
generator while looking at the output of the DAC. Record the maximum frequency that you can recognizably reproduce. Remove the filter
capacitor and estimate the sampling frequency by counting the number
of steps per division.

　　　Maximum Frequency = _____ **Hz**

　　　Sampling Frequency = _____ **Hz**

QUESTIONS:

1. What would have been the resolution of the ADC in step 4 if only 6 bits
were used in the encoding and decoding schemes? Explain why.

2. What would the LEDs display if you didn't include the DC offset on the
input signal in step 5?

3. If only 6 bits were used in the encoding and decoding schemes, would
the DAC output signal appear more or less distorted in steps 2 and 3?
Explain why.

4. How important is it that the clock rate be set high in the encoding of the
digital data? What are the limitations of the clock?

5. What are the limitations of using a low pass filter to shape the output of
the DAC?

PULSE CODE MODULATION AND SERIAL DATA PROTOCOLS

OBJECTIVES:

    1. Demonstrate pulse code modulation (PCM) of an analog signal.

    2. Demonstrate converting PCM data to the standard ASCII serial format.

    3. Use a serial peripheral interface (SPI) digital-to-analog converter (DAC) to turn the PCM signal into an analog signal.

REFERENCE:

Refer to Sections 7-4 and 7-5 of the text.

EQUIPMENT:

Dual-trace oscilloscope
Low-voltage power supply
Function generator
Digital multimeter
Breadboard

COMPONENTS:

Arduino microcontroller
Multi-turn 10-kΩ potentiometer
100-nF capacitor (2 each)
1.5-kΩ resistor
MCP49x1 DAC
Pushbutton switches (2 each)

INTRODUCTION:

Pulse-code modulation (PCM) is a technique to create a digital representation of sampled analog signals. In order to encode the PCM stream, an analog-to-digital converter (ADC) must sample the magnitude of the analog signal at regular intervals. The limitations of the digital representation are based on two parameters: the number of bits used to encode each analog value and the frequency at which the signal is sampled. PCM encoding is often described by a pair of numbers such as "8 bit, 22 kHz," that determine the ability of the PCM encoding to reproduce the original signal. The term 8 bit means that 8 bits are used to encode each analog value, which would give a quantization of 5 V/256 or approximately 20mV. The data rate of 22 kHz means that the encoding will be theoretically limited to signals that are 11 kHz, and in practice limited to signals at even lower frequency.

In the first part of this experiment an analog signal is converted to digital using the onboard ADC in an Arduino microcontroller. This signal is first encoded in a straightforward PCM format and then converted into the ASCII format. The ASCII format is an international standard for encoding characters using 8 data bits.

In the second part of the experiment an external digital-to-analog converter (DAC) is used to take the PCM data from the first part of the experiment and to convert it back into an analog signal. This DAC implements the Serial Peripheral Interface standard, which uses three wires to transmit PCM data from the microcontroller to the DAC.

PROCEDURE:

Part I: PCM and ASCII

1. You will be using the Arduino microcontroller. A microcontroller is a small computer with programmable input and output. In this case you will use a program to have the Arduino act as an 8-bit serial ADC with a custom PCM output. Build the circuit shown in Figure 17-1 below. Be sure to connect the +5V and the ground (GND) connections from the Arduino to your breadboard.

2. If your Arduino is not preprogrammed, open up the Arduino IDE on your computer and load the program "ArduinoPCM." Press the compile and download button to program your microcontroller.

3. Connect your oscilloscope to D4 on the Arduino. Adjust the potentiometer so that there is there is 0 V going to A0. You should observe a series of two pulses that repeat. Sketch the pattern on the oscilloscope in the space below.

4. Adjust the potentiometer so that there is 5 V going to port A0. You should now observe a series of repeating pulses. Sketch the pattern on the oscilloscope in the space below.

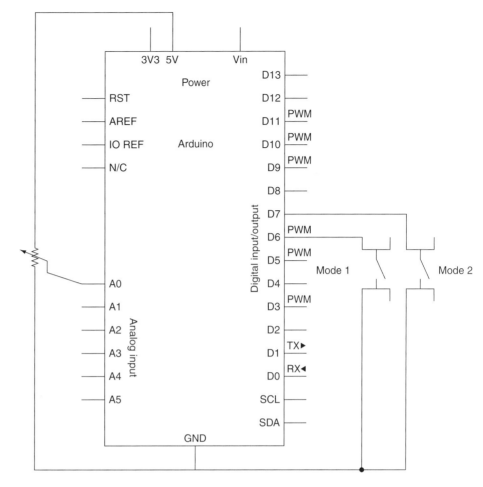

FIGURE 17-1 8-bit serial ADC.

5. The serial PCM protocol in use includes a single start bit and a single stop bit. These are the only bits visible when the input to A0 is 0 V and these bits are longer than the data bits. Start bits and stop bits help determine when to start and stop reading a digital word. Measure the pulse width of the start and stop bits.

Start Bit Pulse Width = _____ *μ*s

Stop Bit Pulse Width = _____ *μ*s

6. This PCM protocol uses the convention that 5-V pulses correspond to a digital 1 and 0-V pulses correspond to a digital 0. Measure the pulse width of one of the 5-V data pulses and the period of the individual data pulses. This period should be similar to the start and stop bit pulse widths.

High Data Bit Pulse Width = _____ *μ*s

Data Pulse Period = _____ *μ*s

7. A single PCM word consists of 11 bits: the *start* bit, followed by 9 *data bits*, followed by a *stop* bit. This demonstration program builds in a slight delay after the sending of the stop bit to make it easier to see the data. Measure the period between PCM words. The data rate in words per second is the inverse of the period. Calculate the data rate in words per second. Data rates are often cited in units of baud, which is bits per second. Calculate the data rate in baud.

Period between PCM words = _____ **ms**

Data Rate = _____ **words/s**

Data Rate = _____ **baud**

8. With 9 data bits, the number of possible states is 2^9, or 512. The voltage is being encoded directly into the bits in units of centivolts (1/100 of a volt). A voltage of 3 V would then be 300 cV or encoded as 300; 300 in binary is 100101100. Adjust your potentiometer to produce 3 V and sketch the pattern of a single word from your oscilloscope in the space below.

9. Here the PCM is presented in a deliberately human readable form. However, this human readable form is out of order from a time perspective. The bit that appears first in time is the least significant bit of the data stream (i.e., the smallest value). In a standard PCM format, the data bits are sent in order from least-significant bit (lsb) to most-significant bit (msb). Pressing the "Mode 2" button will change the time order to the standard machine readable format. Press the "Mode 2" button and sketch the pattern of a single word associated with a 3-V input in the space below.

10. The PCM data you have observed is a custom protocol that uses 11 bits to represent a single analog voltage. However, there are international standards for representing serial data such as the ASCII standard. The most common implementation of the ASCII standard uses 8 data bits, 1 start bit, and 1 stop bit and is often referred to as 8N1. Figure 17-2 shows a "0" encoded in ASCII.

11. Move your oscilloscope probe to Arduino PIN 1 (TX). By pressing the "Mode 1" button, you can observe a TTL (0 is 0 V and 1 is +5 V) ASCII signal of the voltage in units of volts. Set the potentiometer to produce ~1 V and sketch the pattern that you observe in the space below.

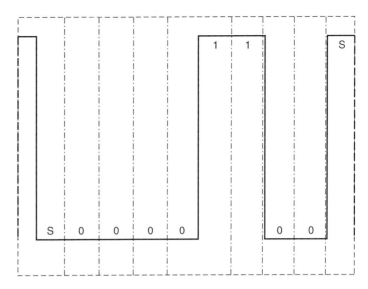

FIGURE 17-2 ASCII for "0" is read right to left as 00110000. This corresponds to a decimal value of 48.

12. Adjust the potentiometer and use your scope to observe the patterns for 0-5 V. Use these patterns to complete Table 17-1.

TABLE 17-1   Binary Patterns for Digits 0 Through 9.

NUMBER	DECIMAL	BINARY
0	48	00110000
1	49	00110001
2	50	
3	51	
4	52	
5	53	
6	54	00110110
7	55	00110111
8	56	00111000
9	57	00111001

13. When the voltage is in centivolts, more than one value must be transmitted. In this case the ASCII protocol just separates the individual values by start and stop bits. Set your potentiometer to 2.25 volts and sketch the resulting output signal in the space below without pressing any buttons.

14. Now the dc input voltage to the ADC will be replaced with an analog signal. Disconnect the wire to the potentiometer and replace it with the function generator signal. Adjust the function generator to produce a 2-V peak-to-peak sine wave centered at 1 V at a very low frequency of 0.1 Hz. You should see the ASCII signals change as the input oscillates. Slowly increase the amplitude of the sine wave until it causes the ASCII code to go from a minimum of 00110000 00110000 to a maximum of 00110101 00110001 00110001, to represent the negative and positive peaks of the sine wave, respectively. Measure the amplitude of the voltage (V_{in}) that causes this to happen.

$$V_{in} \text{ (max)} = \underline{\hspace{2cm}} \text{Vp-p dc}$$

Part II: Digital-to-Analog Conversion (DAC)

15. The same Arduino microcontroller and program as in Part I will be used. A Microchip MCP49x1 digital-to-analog converter (DAC) will be connected to the Arduino to generate an analog signal from the PCM signal. Build the circuit shown in Figure 17-3. (The 12-bit MCP4921, 10-bit 4911 or 8-bit 4901 DAC can be used. The value for bit rate needs to be adjusted in the Arduino code accordingly.)

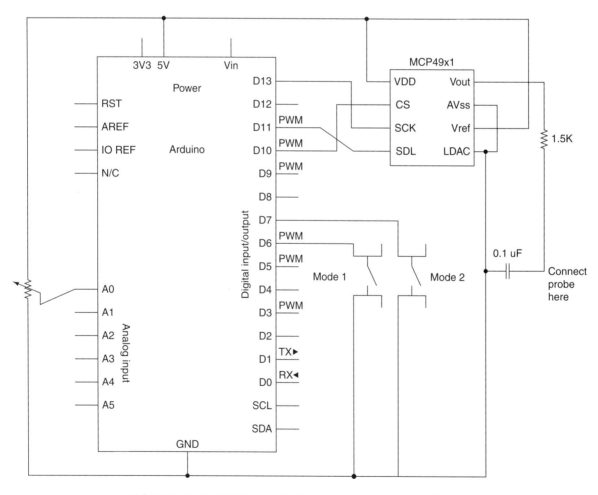

FIGURE 17-3 MCP49x1 digital-to-analog converter (DAC).

16. Test the DAC by connecting the oscilloscope probe to the output point shown in Figure 17-3 to your oscilloscope. Observe that if both the MODE 1 and MODE 2 buttons are pressed simultaneously while the potentiometer is adjusted, the output voltage changes. Adjust the potentiometer to the central position and measure the input to A0 with your digital multimeter. Measure the output of the DAC while both buttons are pressed using the oscilloscope connected at the output point shown in Figure 17-3. Record these values below.

V_{in} = _____ V dc

V_{DAC} = _____ V dc

17. The DAC that is being used is itself a SERIAL device using the Serial Peripheral Interface (SPI) bus. This implementation of the SPI bus uses three wires on pins 10, 11, and 13 of your Arduino (see Table 17-2 below).

TABLE 17-2   SPI Functions and Associated Pins on DAC.

PIN	NAME	FUNCTION
10	CS	Start Transmission
11	SDI	Serial Data In
13	CLK	Clock

Each transmission starts by setting the CS pin low. Connect your oscilloscope to the CS pin (10) and measure the period while pressing the "Mode 1" button. Measure the period of the CS pin while pressing both the "Mode 1" and "Mode 2" buttons.

$T_{CS\,1}$ = _____ μs

$T_{CS\,1\&2}$ = _____ μs

18. While the CS pin is low, data can be sent to the DAC through the serial bus. To send data, a bit is set on the SDI pin, and the clock pin is pulsed. This is repeated 16 times to send two 8-bit bytes consisting of the PCM data to be converted into an analog signal. You will notice a brief space between the two bytes that make up the 16-bit word. Connect your oscilloscope to the CLK pin (13) and measure the period while pressing the "Mode 1" button. Measure the period of the CLK pin while pressing both the "Mode 1" and "Mode 2" buttons. Calculate the frequency of the CLK signal for the two modes.

$f_{CLK\,1}$ = _____ kHz

$f_{CS\,1\&2}$ = _____ kHz

19. Connect your oscilloscope to the SDI pin. The frequency of the signal at the SDI pin is expected to be the same as that of the clock pin. The SDI pin will contain the output PCM stream that is generated from the Arduino and sent to the DAC. This stream consists of 4 configuration bits and 8, 10, or 12 data bits depending on model (8 bits for 4901, 10 bits for 4911, and 12 bits for 4921). The 4 configuration bits are currently set to 0011, and an explanation of these configuration bits is available in the MCP49x1 datasheet. With your potentiometer set to ~0 V, sketch the pattern associated with the SDI pin while pressing the "Mode 1" button.

20. If you adjust your potentiometer while pressing "Mode 1" until the input is 2 V, you should obtain a trace like that shown in Figure 17-4 (shown for a 12-bit DAC).

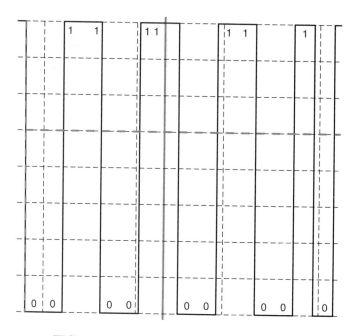

FIGURE 17-4 16-bit word in PCM format.

These pulses can be understood as follows: The first 4 bits are the configuration 0011. The next 12 bits are the data 0011-001100101. The dash corresponds to the vertical line in the figure, which represents the pause between the two bytes that make up the word. Converting 0011-00100101 to decimal gives a value of 1637. Therefore 2V corresponds to the 1637 intervals of the DAC. Calculate the resolution of the DAC as the ratio of the voltage to the number of intervals.

Resolution = _____ mV/Interval

21. Adjust your potentiometer to ~2 V and sketch the pattern associated with the SDI pin while pressing both the "Mode 1" and "Mode 2" buttons. The PCM format used is left to right on the screen just as in the pattern we developed in step 8.

22. Apply a 20-Hz, 2-V p-p signal to A0. Pressing the "Mode 1" button will generate a DAC signal at a low sampling rate of 10 Hz. Observe the signal at the function input on channel A of your oscilloscope and the resulting DAC output (pin 4) on channel B. Let the input signal serve as the trigger for the oscilloscope. Sketch the signal of your DAC output and your function input on the graph below.

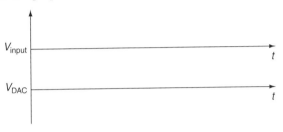

23. Using the same input as in the previous part, press and hold both the "Mode 1" and "Mode 2" buttons to generate a DAC signal at a higher sampling rate of 5 kHz. Observe the signal at the function input on channel A of your oscilloscope and the resulting DAC output (pin 4) on channel B. Sketch the signal of your DAC output and your function input on the graph below.

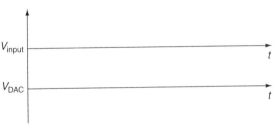

24. While pressing both the "Mode 1" and "Mode 2" buttons, observe what happens when V_{in} is changed to a 100-Hz triangle wave and a 100-Hz square wave and sketch the results in the space below.

25. Switch back to a sine wave and increase the frequency of the signal generator while looking at the output of the DAC. Record the maximum

frequency that you can recognizably reproduce while pressing both the "Mode 1" and "Mode 2" buttons. Remove the filter capacitor and estimate the sampling frequency by counting the number of steps per division.

Maximum Frequency = _____ Hz

Sampling Frequency = _____ Hz

QUESTIONS:

1. Why are STOP and START bits necessary in a digital protocol? Describe a case where they both may not be required.
2. Draw the TTL ASCII representation for the number 7. Label your diagram and explain how it is decoded.
3. Draw the TTL ASCII representation for the number 23. Label your diagram and explain how it is decoded. Explain the bit order of the two numbers.
4. Calculate the frequency associated with the CS pin when the "Mode 1" and "Mode 2" buttons were pressed. How does this frequency compare to the maximum frequency and sampling frequency that you determined in step 11?
5. Was the serial data measured in steps 6-9 characterized as being MSB first or LSB first? Why?
6. How important is it that the clock rate is set high in the encoding of the digital data? What are the limitations of the clock?

FREQUENCY SHIFT KEYING MODULATION AND DEMODULATION

OBJECTIVES:

1. To become familiar with modems.
2. To test and evaluate a simple frequency-shift keying (FSK) modulator based on a 555 timer.
3. To study the operation of an FSK demodulator and use it to recover an encoded signal.
4. To develop a minimum-shift keying (MSK) implementation of an FSK communication system.

REFERENCE:

Refer to Section 8-1 in the text.

EQUIPMENT:

Dual-trace oscilloscope
Low-voltage power supply
Function generator
Multimeter
Breadboard

COMPONENTS:

555 timer integrated circuit
50-kΩ potentiometer
2 each 100-nF and 10-μF capacitors
1 each 220-Ω, 39-kΩ, and 47-kΩ resistors
2N3906 PNP transistor or equivalent

One means by which digital signals are impressed onto an RF carrier is through a modified form of frequency modulation known as *frequency-shift keying* (FSK). Early forms of radio-teletype used such a form of modulation. Today, this digital modulation technique is generally obsolete in its basic form, although the general principles of FSK are used in more advanced data-encoding techniques. However, several third-party vendors have used FSK recently to interface low-speed sensors with iPhone/Android phones using the microphone jack.

In FSK, the two digital logic states, "1" and "0," are converted to a constant-amplitude sine wave that is shifted between two possible frequencies. These two frequencies are referred to as the "mark" and "space" frequencies. These frequencies are usually in the audio-frequency spectrum. Popular mark/space frequency pairs are 1070/1270 Hz, 2025/2225 Hz, and 2125/2290 Hz. For example, the mark frequency of 2025 Hz might represent the binary "1" and the space frequency of 2225 Hz might represent the binary "0." In a radio transmitter, if an FSK signal is fed into the microphone input and single-sideband modulation is used, the RF carrier at the output of the transmitter will then be shifted between the corresponding RF mark and space frequencies.

Note that in the FSK encoder system shown in Figure 18-1, the resulting SSB output signal ends up being an elementary form of FM, since only a single sine wave actually modulates the RF carrier at a time. Being FM makes it quite immune to noise interference. This is the main advantage of using FSK in a digital communications system. As seen in Figure 18-2, the SSB receiver would detect the original audio frequencies and the FSK decoder would then convert them back to the original digital format.

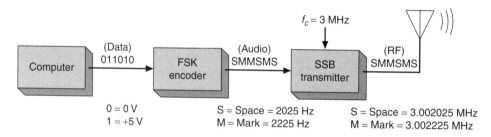

FIGURE 18-1 FSK encoder and transmitter.

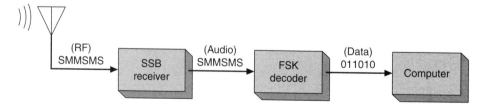

FIGURE 18-2 Digital communication receiver and decoder.

In some digital communications systems, regardless if radio or telephone wires are being used to send the FSK signals, there is a need for two-way communications to occur simultaneously. Thus the FSK encoding and decoding are needed at both ends of the communications link. If this is being done, "modems" are used (see Figure 18-3).

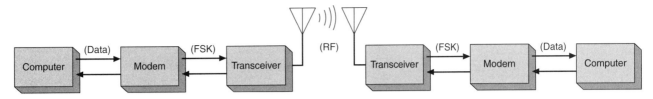

FIGURE 18-3 FSK system with modems in transmit and receive paths.

"Mo-dem" is an acronym for a device that contains an FSK encoder or *MOdulator* and an FSK decoder or *DEModulator*. In this laboratory exercise, an FSK modulator is built from a 555 timer and an FSK demodulator is implemented using an Arduino UNO.

PROCEDURE:

Part I: FSK Modulation

1. Using the datasheet for the 555 timer operating as an ASTABLE oscillator, determine the possible range of output frequencies for the square wave signal produced by the oscillator in Figure 18-4.

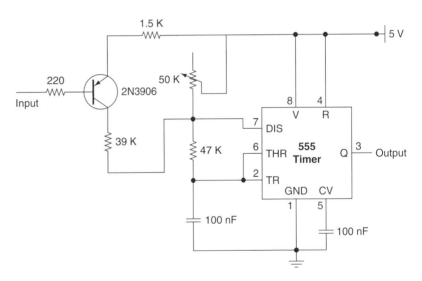

FIGURE 18-4 FSK modulator circuit.

2. Build the 555 oscillator following the schematic shown in Figure 18-4. In this circuit your input signal will be a square wave that turns on and off the PNP transistor shown on the left side of the schematic. This will change the voltage seen at pin 7 and thus change the frequency of the 555 oscillator.

3. Ground the input to the transistor and apply 5 V to the circuit. Observe the output pulse of your 555 timer on the oscilloscope. Adjust the potentiometer until there is an approximately 60% duty cycle and a 1–1.2-kHz

output. This signal will serve as the mark frequency. Measure its frequency and duty cycle.

Mark Frequency _____ Hz   Duty cycle _____ %

4. Now connect the input to the transistor to 5 V and observe the output of your 555 timer. This is the space frequency of the FSK modulator. Record the frequency and the duty cycle of the space frequency in the space below.

Space frequency _____ Hz   Duty cycle _____ %

5. Calcuate the center frequency as the average of your mark frequency and your space frequency.

Center Frequency _____ Hz

6. Now apply a 100-Hz, 5-V square wave centered at 2.5 V to the input. This will serve as your data signal. Connect this to your oscilloscope input A and the output of your 555 to input B. Trigger on input A and sketch the output in Figure 18-5 below. You should observe the output waveform switching between the mark and space frequencies. The square-wave generator is simulating a computer sending digital data. The voltage you measure is the encoded FSK signal.

FIGURE 18-5 Input and frequency-shift modulated output signals from 555 timer.

7. Increase the frequency of the square wave while watching the oscilloscope display. At some point, you should cease to observe the separate mark and space frequencies. Record the frequency at which you no longer observe separate mark and space signals. What is the frequency of the signal that you now observe?

Cutoff frequency _____ Hz   Observed frequency _____ Hz

Part II: FSK Demodulation

8. Add the Arduino UNO to the FSK modulator circuit you built in part 1 as shown in the schematic of Figure 18-6. Be sure to remove the external power supply as the circuit will now be powered from the Arduino. The Arduino will serve to demodulate the FSK signal.

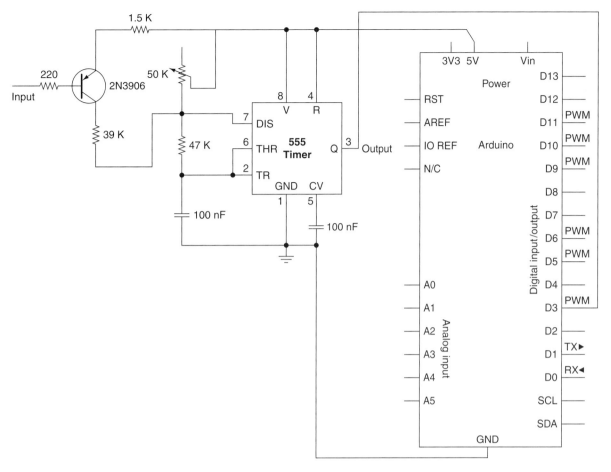

FIGURE 18-6 FSK modem.

9. If your Arduino is not preprogrammed, open up the Arduino IDE on your computer and load the program "ArduinoFSK". Press the compile and download button to program your microcontroller.

10. When the Arduino UNO is first powered, it measures the incoming FSK signal and determines the MARK, SPACE, and CENTER frequencies just as was done in Part I of the lab. Apply a 100-Hz, 5-V square wave centered at 2.5 V to the input. This will serve as your data signal. Connect this to your oscilloscope input A and the output of your 555 to input B. Verify that your signal has not changed by recording the MARK, SPACE, and CENTER frequencies.

Space Frequency _____Hz   Mark Frequency _____ Hz

Center Frequency _____ Hz

11. Now connect channel B of your oscilloscope to output 13 on the Arduino, which is the FSK demodulation output. Sketch the data signal and the demodulated signal in Figure 18-7.

FIGURE 18-7 Data and demodulated signals from Arduino demodulator.

12. Adjust the data frequency of the input to your oscillator and observe the changes in the demodulation output. Turn the frequency of the input up until you reach the cutoff frequency measured in step 7. Sketch the output of the demodulator at the cutoff frequency in Figure 18-8 below.

FIGURE 18-8 Data and demodulated signals at cutoff frequency.

13. A widely used variation on FSK is known as *minimum-shift keying* (MSK). You will have noted that sometimes the mark or space is not the full width you measured in steps 3 and 4, due to the signal shifting in the middle of a mark or space. In MSK the periods of the mark and space frequences are such that cycles of the sine-wave carier cross zero right at the edges of the modulating signal pulse transitions. A simple way to achieve this is to set the difference between the mark and space frequency to be equal to one-half the lower frequncy. Adjust your mark and space frequencies to meet this criteria by adjusting the 50-kΩ potentiometer. Record your new mark, space, and center frequencies below:

Space frequency _____ Hz   Mark frequency _____ Hz

Center frequency _____ Hz

14. Now apply a 100-Hz 5-V square wave centered at 2.5 V to the input. This will serve as your data signal. Connect this to your oscilloscope input A and the output of your 555 to input B. Trigger on input A and sketch the output in Figure 18-9 below.

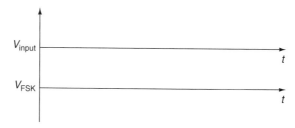

FIGURE 18-9 Input and output waveforms for minimum-shift keyed signals.

15. Increase the frequency of the square wave while watching the oscilloscope display. At some point, you should cease to observe the separate mark and space frequencies. Record the frequency at which your no longer observe separate mark and space signals. What is the frequency of the signal that you now observe?

Cutoff Frequency _____ Hz

Observed Frequency _____ Hz

16. Now connect channel B of your oscilloscope to output 13 on the Arduino, which is the FSK demodulation output. Sketch the data signal and the demodulated signal in Figure 18-10 below.

FIGURE 18-10 Input and demodulated output waveforms for MSK signals.

QUESTIONS:

1. Calculate the maximum baud rate that the digital communication link can handle without any errors resulting.

2. In your own words, write a brief theory of operation section for the digital communications link analyzed in this experiment. Assume that this system is being sold as a product to the technical public and that this theory section is to be incorporated as part of a manual for the product.

3. How is the cutoff frequency you measured related to the mark and space frequency?

4. What does the demodulator circuit do when the input frequency exceeds the cutoff (or bit clock)? Explain in detail. Is this behavior always the same?

5. In the MSK operating mode, how did the cutoff frequency compare to the mark and space frequencies? How did this cutoff frequency compare to the cutoff frequency in the non-MSK implementation in Part I?

6. Write a technical test procedure for the calibration of the FSK encoder for the generation of MSK.

MODEM COMMUNICATIONS

OBJECTIVES:

1. To explore modem communications between two personal computers.
2. To become familiar with the AT command set.
3. To demonstrate low speed serial communications using a variety of protocols.

REFERENCE:

Refer to Section 2 in the chapter "Telephone Networks," Sections 2 and 8 in the chapter "Computer Communication and the Internet," and Section 2 in the chapter "Transmission Lines" Sections 9-2, 11-2, 11-8, and 12-2 in the text.

EQUIPMENT:

Two personal computers with internal or external modems installed

dc power supply

Breadboard

Multimeter with test leads

RJ11 or RJ45 crimping tool

COMPONENTS:

380-Ω ½-W resistor

Capacitors: 1 each 1 μF 10 μF, and 47 μF

CAT5 Keystone jack

RJ11 connectors

INTRODUCTION:

Most home users access the internet with broadband connections such as high-speed cable modems, digital subscriber lines, or fiber-optic-based technologies. However, conventional "dial-up" internet access through analog phone lines and the Public Switched Telephone Network (PSTN) is alive and well even today. In the early days of the internet, America Online blanketed the country with compact discs containing the setup files home users needed to become AOL subscribers. Perhaps surprisingly, and even with the widespread availability of broadband service, AOL has approximately 3.5 million dial-up internet users. AOL and other internet providers still offer dial-up internet connectivity because broadband technologies and services are not available in rural areas and because not all users need, or are willing to pay for, high-speed service. Clearly, then, modem-based technologies are still relevant even in an era where broadband connectivity is considered nearly universally available.

Dial-up internet access, and computer communication through analog phone lines generally, is achieved with devices called modems. The term is an acronym for *MO*dulator/*DEM*odulator. The modem converts digital signals from a computer into analog signals that can be carried across phone lines to a distant computer, at which point a modem at that location converts the analog signal back to a digital form suitable for use by the distant computer. The three main parts of a modem are the microcontroller unit (MCU), the data pump unit (DPU), and the data access arrangement (DAA).

In this lab, we will use either internal (i.e., a circuit board installed on or built into a computer motherboard) or external modems and other switching network connections to demonstrate the fundamentals of computer-to-computer communications. The modems we will use are said to be *Hayes compatible* because they make use of instructions originally developed for a product introduced in 1981 called the Hayes Smartmodem. These instructions collectively form the *AT command set,* which has become an industry standard and is still the most widely used set of commands for modems from all vendors. Our goal will be to have two computers communicate with each other through a simulated phone line using the modems and a software terminal emulator program called Tera Term.

Like other equipment designed to interface with an analog telephone line, modems respond to the "off-hook" condition by sensing current flow within the line. In this lab, we will build a dial tone generator, shown in Figure 19-1, to simulate the conditions of an analog phone line by applying a current of approximately 20 mA to the cable used to connect the two computer modems in the laboratory. This current will cause the modems to respond as though they are connected to an actual dial-up line, even as analog lines are becoming increasingly rare in today's digital world. The capacitors in the circuit are designed to filter out any unwanted signals.

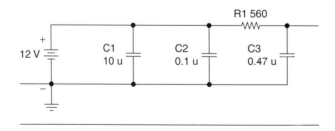

FIGURE 19-1 Phone line simulator circuit.

Data communications protocols are used for communications between two (or more) pieces of equipment. One end is designated as the data terminal equipment (DTE) and the other is called the data communications equipment (DCE). A desktop computer is an example of a DTE because it prepares data to be transmitted. A computer connected to a modem transfers data at the DTE rate. The modem is the DCE and its speed determines the DCE communications rate. The terminal end of the connection is the DTE, while the modem itself is the communication end or DCE. In network communications the clock signal is provided by the DCE and the DTE device synchronizes to this clock signal.

PROCEDURE:

1. Ensure that the computer is configured with a working Microsoft Windows operating system (Any version XP, Vista, 7) with an installed internal or external modem. Ask your instructor for assistance.

2. Build the dial tone generator circuit shown in Figure 19-1. Have your instructor inspect your completed circuit before moving to the next step.

3. Obtain or construct a phone line cable to connect your workstation's modem to the phone line circuit. Use RJ11 connectors and a crimping tool to construct a cable by following these steps:

   a. From a spool of phone line, cut off a piece of cable long enough to connect between your computer and the generator circuit.

   b. Using the crimping tool, strip off approximately ½" of the outer insulator from the phone line cable.

   c. Slide the RJ11 connector over the four wires (Black, Red, Green, Yellow).

   d. Insert the connector and cable into the crimper and squeeze the handle.

   e. Repeat on the other end of the cable.

4. On your workstation, download and install Tera Term (http://logmett.com/index.php?/download/tera-term-476-freeware.html)

5. Open Tera Term and connect to the serial port on which your modem is installed.

6. Type the following commands in the terminal.

   a. AT

   b. ATI

   c. ATZ

   d. ATX3&C0

7. You should see an "OK" back on the screen. If not, inform your instructor—it is time to troubleshoot! See http://www.computerhope.com/atcom.htm for a complete list of AT commands.

8. Your machine's modem should connect to the phone line circuit on one end and the group across from you should do the same thing. One generator will be used for two computers.

9. In Tera Term click on Setup, Terminal.

10. Click the checkbox for Local Echo, and change the New-Line from CR to CR+LF for both Send and Receive.

11. On one of the workstations type "ATA" to answer the incoming call.

12. On the other workstation type "ATD" to dial the outgoing call.

13. You should see your machines connect. At what speed did your connection negotiate? _____

14. Send a message to the other machine. Have the other machine reply to your message.

15. On your workstation create a text file and name it ModemLab.txt. Type your group's names and today's date and save the file.

16. From within Tera Term, while still connected to the other computer, click on File, Transfer, ZMODEM, Send or Receive.

17. One computer will send the file, while the other computer will receive the file. Transfer and receive the ModemLab.txt file using the default connections. The receiving computer should rename the file and open it when the transfer is complete. At what speed is the file being transmitted? _____ _____

18. Reverse the procedure, so if you sent the file now you will receive it. Where is the file located? (Give complete path.) _____ _____

19. Repeat the file transfer (send/receive) using another protocol other than ZMODEM. Both sides of the connection must use the same protocol. Which one did you select? _____ _____

20. When you have transferred the file using ZMODEM and at least one other protocol have the instructor initial your lab.

        Instructor's Initials _____

QUESTIONS:

1. In terms of data terminal equipment and data communication equipment (DTE, DCE), identify the following components from this experiment:
   (a) Computer 1 _____
   (b) Modem 1 _____
   (c) Modem 2 _____
   (d) Computer 2 _____

2. What does the following command do? ATX3&C0 (that is a zero)? _____ _____

3. Why was this command necessary for our setup? _____ _____ _____

4. What is the maximum speed that two computers with modems can communicate? _____ _____

5. List the file transfer protocols that Tera Term supports. _____ _____ _____ _____

NAME _____

ROUTER CONFIGURATION

OBJECTIVES:

1. To explore data communications between two endpoints using wide-area network (WAN) technologies.

2. To become acquainted with the command line interface used to program Cisco routers.

3. To demonstrate high-speed serial communications between endpoints on a WAN.

4. Use the Ping utility to verify connectivity between networked devices.

REFERENCE:

Refer to Sections 9-4, 9-5, 11-4, and 11-6 of the text.

EQUIPMENT:

Two personal computers with internal or external serial ports; USB to serial adapters will also be suitable

Two Cisco routers with fast Ethernet port and a serial WAN interface card (WIC-1T or 2T)

One Cisco smart serial cable compatible with the WIC-1T connection on the routers

COMPONENTS:

Tera Term Software
(http://download.cnet.com/Tera-Term/3000-20432_4-75766675.html)

INTRODUCTION:

The internet backbone is composed of thousands and thousands of *routers*, devices that interconnect various computer networks. Most home internet users connect to their internet Service Providers (ISPs) with high-speed broadband connections such as DSL, cable, or fiber-optic line. In each case the ISP provides home users with routers for connecting their computers to the internet. In an enterprise network, such as for a business, government agency, or school, information-technology professionals will install routers from well-known telecommunications equipment manufacturers such as Cisco, Dell, and Hewlett-Packard. Arguably, the most commonly used routers in enterprise networks are from Cisco Systems. For this reason, Cisco routers will be the focus of this experiment, although the underlying principles can be applied to equipment from other vendors as well.

In a high-speed connection there are two endpoints: the data terminal equipment (DTE—your home computer) and the data communication equipment (DCE—your ISP's router). The DCE sets the clock rate and bills you accordingly. The communication between these endpoints is carried on a high-speed communication circuit such as frame relay, T-1, DS-3, DSL, cable, or fiber-optic line. A T-1 channel is a time-division multiplexed (TDM) data stream at a constant rate of 1.544 Mbps. A DS-3 channel is composed of 28 T-1's for an overall data rate of 44.736 Mbps. Frame relay can be thought of as a fractional T-1 in that it can be configured at speeds up to a full T-1.

Cisco routers can be programmed with either a proprietary graphical user interface or a generic command-line interface. In this lab the command-line interface will be used exclusively. The Cisco IOS contains thousands of commands, and a thorough study of Cisco Networking is a multi-year endeavor. For this experiment the necessary commands will be provided.

This lab will incorporate IP addressing; however, a detailed explanation of this vast topic will be covered in another experiment. There are volumes of books that explain the various details of the TCP/IP protocol suite. Again for this experiment the necessary IP addresses and subnet masks will be provided.

Not shown in this experiment, but an essential part of "real world" installations of high-speed data communications networks, is a device called the customer service unit/data service unit (CSU/DSU). Its main purposes are to provide data interface to the carrier (including adding framing information and maintaining data flow), to storing performance data, and to provide line-management functions, including loopback capability for remote troubleshooting purposes.

PROCEDURE:

Part I: Set Router to Factory Settings

The student will be using an excellent freeware terminal program known as Tera Term. This software can be obtained from http://download.cnet.com/Tera-Term/3000-20432_4-75766675.html. It is recommended that all options with the exception of Tera Term be unchecked during the installation.

Before performing this experiment, it is recommended that the routing equipment's existing configuration be erased. This can be accomplished on Cisco routers as follows:

router#**configure terminal**

router(config)#**config-register 0x2102**

router(config)#**end**

router# **write erase**

(do not save when prompted to save the configuration)

router# **reload**

Once the router reloads, the System Configuration Dialog appears. Type in "no" to skip it. Your router's configuration will now be set to factory default settings.

Part II: Network Setup

Figure 20-1 depicts our network setup.

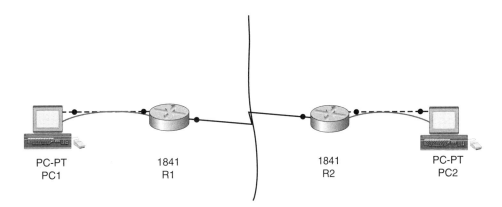

FIGURE 20-1 Network setup.

The equipment on the left side will be considered the customer's, and the equipment on the right side will be considered the ISP's. Table 20-1 shows the connections between the equipment.

TABLE 20-1 Connections between Customer and ISP Equipment.

DEVICE	DEVICE	CONNECTION	CABLE
PC1 (RS232)	R1 (Console)	Console	Roll Over
PC2 (RS232)	R2 (Console)	Console	Roll Over
PC1 (Fast Ethernet)	R1 (Fast Ethernet)	Cat 5e	Crossover
PC2 (Fast Ethernet)	R2 (Fast Ethernet)	Cat 5e	Crossover
R1 (Serial)	R2 (Serial)	Serial	Cisco Smart

Table 20-2 shows the IP addressing we will use for this experiment.

TABLE 20-2 IP Addressing Table for PC1 and PC2.

DEVICE	IPv4 ADDRESS	SUBNET MASK
PC1 (Fast Ethernet)	192.168.1.2	255.255.255.0
PC2 (Fast Ethernet)	192.168.2.2	255.255.255.0
R1 (Fast Ethernet)	192.168.1.1	255.255.255.0
R2 (Fast Ethernet)	192.168.2.1	255.255.255.0
R1 (Serial)	10.0.0.1	255.255.255.252
R2 (Serial)	10.0.0.2	255.255.255.252

Note that this lab is performed on two PCs and two routers, so the tasks for PC1 and R1 are shown first, followed by those for PC2 and R2. If this experiment were to be performed by two people, one person could handle the PC1 and R1 tasks and the other the PC2 and R2. Commands in **bold** font are to be typed directly into the router's terminal via Tera Term.

The following steps (1-16) are to be performed on PC1.

1. Set the IP address of PC1 to 192.168.1.2
2. Set the subnet mask of PC1 to 255.255.255.0
3. Set the default gateway of PC1 to 192.168.1.1
4. Connect the roll over cable between PC1's RS232 port and R1's console port
5. Open Tera Term on PC1 and connect to the RS232 serial port used in step 4.
6. Press Enter a couple of times to see the Router's prompt.
7. In Tera Term type **enable**.
8. Type **configure terminal**.
9. Type **interface fastEthernet 0/0**.
10. Type **ip address 192.168.1.1 255.255.255.0**.
11. Type **no shutdown**.
12. Type **exit**.
13. Type **interface serial 0/0/0**.
14. Type **ip address 10.0.0.1 255.255.255.252**.
15. Type **no shutdown**.
16. Type **encapsulation hdlc**.

The following steps (17-32) are to be performed on PC2.

17. Set the IP address of PC2 to 192.168.2.2
18. Set the subnet mask of PC2 to 255.255.255.0
19. Set the default gateway of PC2 to 192.168.2.1
20. Connect the roll over cable between PC2's RS232 port and R2's console port.
21. Open Tera Term on PC2 and connect to the RS232 serial port you connected the roll over cable to.
22. Press Enter a couple of times to see the Router's prompt.
23. In Tera Term type **enable**.
24. Type **configure terminal**.
25. Type **interface fastEthernet 0/0**.
26. Type **ip address 192.168.2.1 255.255.255.0**.
27. Type **no shutdown**.
28. Type **exit**.
29. Type **interface serial 0/0/0**.
30. Type **ip address 10.0.0.2 255.255.255.252**.
31. Type **no shutdown**.
32. Type **exit**.

We will now set the encapsulation protocol and clock rate on R2 (DCE). Remember the DCE provides the clock rate for synchronization.

33. Type **encapsulation hdlc**.

34. Type **clock rate 64000**.

35. Attempt to ping from PC1 to R1's FastEthernet by typing from PC1's command prompt **ping 192.168.1.1**. Was this successful? If not why? _____
_____

36. Attempt to ping from PC1 to R1's Serial interface from PC1's command prompt **ping 10.0.0.1**. Was this successful? If not why? _____
_____

37. Attempt to ping from R1's Tera Term terminal to PC1 **ping 192.168.1.2**. Was this successful? If not why? _____
_____

38. Attempt to ping from PC2 to R2's FastEthernet by typing from PC2's command prompt **ping 192.168.2.1**. Was this successful? If not why? _____
_____

39. Attempt to ping from PC1 to PC2 by typing from PC1's command prompt **ping 192.168.2.2**. Was this successful? If not why? _____
_____

We must tell our routers to advertise their connections so the other routers can dynamically make connections. We will do this using a dynamic routing protocol known as RIP version 2.

Steps 40-48 are to be performed on PC1.
Connect PC1 to R1 via Tera Term:

40. Type **enable**.

41. Type **configure terminal**.

42. Type **router rip**.

43. Type **version 2**.

44. Type **network 192.168.1.0**.

45. Type **network 10.0.0.0**.

46. Type **exit**.

47. Type **exit**.

48. Close PC1's Tera Term connection.

Steps 49-57 are to be performed on PC2.
Connect PC2 to R2 via Tera Term:

49. Type **enable**.

50. Type **configure terminal**.

51. Type **router rip**.

52. Type **version 2**.

53. Type **network 192.168.2.0**.

54. Type **network 10.0.0.0**.

55. Type **exit**.

56. Type **exit**.

57. Close PC2's Tera Term connection.

58. Attempt to ping from PC1 to PC2 again. Was this successful? _____

---

59. If this is not successful, try to ping from PC1 to the following IP addresses:
   a. 192.168.1.1 (Success / Failure)
   b. 10.0.0.1 (Success / Failure)
   c. 10.0.0.2 (Success / Failure)
   d. 192.168.2.1 (Success / Failure)

60. If you are using a Windows PC your ping from PC1 to PC2 may be blocked by the software firewall of the operating system. Temporarily disable the firewall and try to ping from PC1 to PC2 again.

QUESTIONS:

1. What is the purpose of the clock rate? What device establishes the clock rate?

2. What is the purpose of encapsulation?

3. What is the purpose of a protocol?

4. Where did you see the implementation of a protocol in this experiment?

5. What is the purpose of the CSU/DSU?

6. Where does the CSU/DSU appear with respect to the customer's premise equipment and the carrier's network?

WIRELESS COMPUTER NETWORKS

OBJECTIVES:

    1. To explore wireless networking technology.

    2. To measure the signal strength of various wireless access points (WAPs).

    3. To set up and secure a wireless network using an off-the-shelf WAP.

    4. To set up and secure a wireless network using a power over Ethernet (POE) WAP.

    5. To demonstrate a cellular phone hotspot Wi-Fi network.

REFERENCE:

Refer to Sections 10-1 and 11-3 through 11-5 in the text.

EQUIPMENT:

PC with a wireless network adapter configured

Wireless access point (Linksys WRT54GS or any off the shelf WAP)

Patch cables to connect the WAP to an existing network and to connect a PC to manage the WAP

Power Over Ethernet Switch (Netgear FS726TP or equivalent)

POE WAP (3Com 7760 or equivalent)

Optional: 3.0 GHz spectrum analyzer (Rigol DSA1000 or equivalent)

Optional: SMA reverse jack to N plug adapter

COMPONENTS:

None

INTRODUCTION:

In wireless data communication systems and local-area networks (LANs) the *wireless access point* (WAP) transmits and receives radio-frequency signals to and from computers and other devices in the vicinity of the WAP. The WAP then connects the associated devices to a network or to the internet. Low-power wireless networks are generally known by the term *Wi-Fi* (for "Wireless Fidelity"), but are formally known as IEEE 802.11 networks, a term resulting from the set of LAN/WAN standards developed by working committees of the Institute of Electrical and Electronics Engineers (IEEE). Many varieties of wireless equipment, using different modulation formats and frequency bands and offering varying data rates and operating ranges, share the 802.11 designation. The most widely encountered are the 802.11a, 802.11b, 802.11g, 802.11n, and the newest entry, 802.11ac.

Setting up a home wireless network is not difficult, but one point should be emphasized initially: Securing the network from unauthorized use is an essential part of initial setup. Manufacturers now routinely include steps to ensure network security when their WAPs are first put into service. These steps represent a much-needed change from past years where a consumer could pull a newly purchased WAP out of its box and immediately begin using it without any concern for potential criminal conduct by unauthorized users. Best practice is always to secure a wireless network as one would a wired network.

Effectively securing a wireless access point is accomplished in two steps. The first step is to disable the Service Set Identifier (SSID) broadcast, and the second is to select a security protocol. Disabling the broadcast of the (SSID) prevents the wandering wireless hacker from "seeing" an active network; however, someone with the proper tools could find the network even with the SSID disabled. The next layer involves the selection of a security protocol. Modern equipment offers at least three options here: no security, WEP, and WPA. The "no security" option is just that—no security. The most basic security option, called Wired Equivalent Privacy (WEP), was introduced with the initial 802.11 standards but has proven flawed in many ways. With readily available tools a hacker can "crack" a WEP secured WAP in a matter of minutes. A more robust option, Wi-Fi Protected Access (WPA), was developed to address WEP's weaknesses. WPA has given way to a more secure form, WPA2, which should be chosen if offered as a security protocol.

Wireless equipment for home use is available at nearly all home electronics stores. These off-the-shelf products are designed for the home user. A home WAP is managed by accessing the device through its built-in web browser. Modern home WAPs are very reliable and work well in environments where the number of connected users is relatively small, say fewer than ten. In office environments where there may be more than ten users, a more robust class of WAP should be used. The cost of such devices also increases as the robustness of the system increases.

The network communications professional should be aware of a number of personal computer– and laboratory-based tools. Among the most important is the freeware program inSSIDer (http://www.metageek.net/products/inssider/), which gives users a visual representation of the wireless networks around them. In the laboratory a spectrum analyzer can be used to see wireless signals.

The increased speed of cellular networks has led to the creation of Wi-Fi hotspots that can be set up and shared from a mobile phone. The instructor can demonstrate the setup and operation of such a hotspot if time permits.

PROCEDURE:

1. The following steps make use of the inSSIDer freeware program, which should be installed on a Windows PC equipped with a wireless network adapter. inSSIDer has many signal analysis features, including a spectrum display similar to that produced by a hardware spectrum analyzer.

Follow the steps mentioned next to construct a simple network environment using a home wireless router such as the Linksys WRT54GS. The assumption is that the network will be built in the order presented here; however, this is by no means the only way to accomplish the objectives of the lab.

2. The WAP should be reset to factory defaults. A quick search of the manufacturer's website should provide detailed instructions on how to accomplish this task.

3. Once the WAP is reset, connect the WAP to the existing network by plugging one end of the patch cable into the *internet* port on the WAP and the other end into the wall jack in the laboratory providing internet access.

We will now create a new network segment to use by plugging the power over Ethernet (POE) switch into a LAN port on the Linksys wireless router.

4. Plug the 3Com POE WAP into one of the POE ports on the Netgear switch.

5. Plug your laptop or PC into one of the available Netgear switch ports.

6. Your PC should be set to "Obtain an IP address automatically."

7. We will work with the Linksys WAP first. Your PC will obtain an IP address from the Linksys WAP. Fill in the following information:

   a. IP Address of your PC: _____

   b. Subnet Mask of your PC: _____

   c. Default Gateway: _____

8. You can manage the Linksys WAP by opening an internet browser and navigating to the IP address of the Default Gateway. Logon to the WAP using the default username and password. (For the Linksys WRT54GS the username is "admin" and the password is "admin".)

9. Once logged on perform the following tasks:

   a. Change the SSID to something unique (and easy to remember)

   b. Disable the SSID broadcast

   c. Set the Wireless Security to WEP

   d. Create a WEP key

10. Using your device with a wireless network adapter, attempt to connect to the WAP using WEP and fill in the following information:

    a. Wireless LAN IP Address: _____

    b. Default gateway: _____

11. Next change the Wireless Security to WPA (WPA2 Pre-Shared Key Only if available).

12. Create a WPA key.

13. Using your device with a wireless network adapter attempt to connect to the WAP using WPA or WPA2.

14. We will now bring the POE WAP online by first resetting it to factory defaults. This can be done using a paper clip and holding down the reset button on the back of the WAP for 5-10 seconds and releasing.

15. Once all the lights are on indicating the WAP is ready, go in to the Linksys WAP's **Status, Local Network** settings.

16. Click on DHCP Clients Table and find the IP address of the POE WAP. What is this IP address? _____

17. Open a new internet browser window and navigate to the IP address of the POE WAP.

18. Log on using the default username and password (shown on the logon screen).

19. On the left side menu under System Configuration, click on Wireless Network. We will be editing the Existing Profiles. Be sure to enable the profile you are currently working on. You will want to rename your WAP's SSID from 3Com to something unique.

20. Make sure to enable Multi-BSSD to allow the various profiles to appear.

21. Edit the 3 profiles with the first one set to "No Security," the second one set to "WEP," and the third profile set to "WPA2 Only."

22. From your wireless device attempt to connect to each profile using all three security profiles. Can you connect to all three profiles? If not describe the steps you took to troubleshoot the connection problems.

Our Linksys and 3Com WAPs are broadcasting radio frequencies. We will now use our PC to look at these and determine the signal strength (expressed as the received signal strength indication (RSSI) and expressed in dBm).

23. On your PC with a wireless network adapter and inSSIDer installed, open inSSIDer. Fill in Table 21-1.

TABLE 21-1   Observed Signals, Channels in Use, and Received Signal Strengths for Wireless Network Adaptor.

SSID	CHANNEL	RSSI (dBm)	MAX RATE

24. Ensure that the Linksys and 3Com WAPs are set to different channels.

25. Click on the **2.4 GHz Channels** tab.

26. We will now investigate the effect of moving the WAP various distances from the PC running inSSIDer. (Depending on your setup you could also move the PC running inSSIDer). Set the distance between the PC and the two WAPs according to Table 21-2 and record the amplitude in dBm.

The following procedures can be done as a demonstration. You will need a smartphone that is capable of creating a Wi-Fi hotspot.

27. Create a cellular smart phone Wi-Fi hotspot and connect your wireless device to this network. The type of phone will dictate the instructions for this procedure.

28. What IP address did your wireless device obtain from this hotspot connection?

29. Repeat step 34 for this wireless connection. How do your results compare?

TABLE 21-2 Effect of Distance on Signal Strength.

WAP NAME	DISTANCE APART	DBm
Linksys	1 meter	
Linksys	2 meter	
Linksys	4 meter	
3Com	1 meter	
3Com	2 meter	
3Com	4 meter	

The following procedures can be done as a demonstration. You will need a spectrum analyzer capable of operation at 2.5 GHz and an appropriate antenna. Some commercial WAPs have detachable antennas that are suitable for this demonstration. A reverse-polarity SMA to N plug adapter must be used to connect this antenna to the spectrum analyzer.

The following procedure uses the 3Com WAP set to 802.11b with a 1 Mbps data rate to view the characteristics of its RF signal on the spectrum analyzer.

30. Log on to the website for the 3Com WAP and under Wireless Network select "802.11b" for the "Wireless Mode."

31. Click on "Wireless Advanced Settings" and change the Data Rate to 1 Mb.

32. Connect your PC to this wireless network and start a large download.

33. On the spectrum analyzer set the center frequency to the frequency of the channel being used by the WAP. (See Table 10-1 in the text.)

34. Set the frequency span for 40 MHz.

35. Set analyzer trace for maximum hold.

36. After about 60 seconds the frequency spectrum displayed should be similar to that of Figure 21-1.

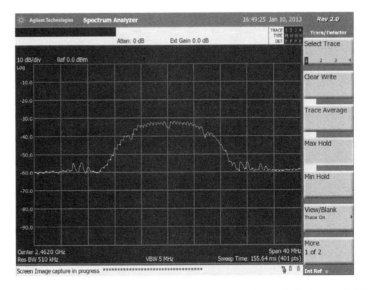

FIGURE 21-1 Frequency spectrum of 802.11b modulation at 1 Mb.

37. With a frequency span of 40 MHz, each major division on the frequency axis represents one-tenth the frequency span, or 4 MHz. Referencing Figure 21-1, what is the occupied bandwidth? How closely does this measurement agree with the expected bandwidth per channel for IEEE 802.11b emissions as described in Section 10-1 in the text? What kind of transmission is this?

QUESTIONS:

1. What are the most common wireless network security protocols available on modern WAPs?

2. How many hexadecimal characters must be used for 64-bit wired equivalent privacy (WEP) encryption?

3. What is the relationship between the loss in dB and the distance between the devices?

4. What might account for any discrepancies between the measured results and the predicted relationship between the loss in dB and the distance between the devices?

5. Why does the graphical representation of the radio signal in inSSIDer look the way it does? In other words, what type of modulation is being represented?

6. Why is it recommended that WAPs be configured for channels 1, 6, or 11?

7. Would two (or more) WAPs that are located close to each other and configured for the same channel work properly? Why or why not? What are the possible drawbacks of such an arrangement?

NAME _____

PLANNING AND DESIGNING LOCAL-AREA NETWORKS (LANs)

OBJECTIVES:

1. To design a logical network diagram based on a client interview.

2. To create a network diagram using network diagramming software.

3. To build a physical network using fast Ethernet networking equipment.

4. To construct a straight-through network cable.

5. To construct a cross over network cable.

REFERENCE:

Refer to Sections 11-3 through 11-5 in the text.

EQUIPMENT:

Two personal computers with functioning network interface cards (NIC)

Two multimode fiber-optic to 10/100 Mbps Ethernet media converters

One ST or SC multimode patch cable

One fast Ethernet switch

Category 5e stranded UTP cable in bulk

COMPONENTS:

CAT5 Keystone jack, RJ45 cable ends, networking crimping tool

INTRODUCTION:

In this experiment, you will produce a conceptual design for a local-area network (LAN) based on a set of user requirements for a hypothetical organization. The steps to be followed illustrate the general approach to building a network from the ground up, which can appear to be a daunting task even for the most experienced networking professional. A substantial portion of the effort involves ascertaining user requirements precisely, developing topology diagrams to establish both the physical locations of equipment as well as the signal flows among system components, and, finally, constructing the network and documenting the "as-built" configuration. Each phase will be covered in the steps to follow. With proper planning and design, the network deployment process can be transformed from "daunting" to "doable."

The importance of thorough, up-to-date documentation cannot be overstressed. For this reason, a substantial part of this experiment is devoted to a review of the essential elements of well-written documentation. The perhaps natural reluctance to give adequate attention to the documentation "chore" seems understandable at first. Network engineers are under time constraints, and once a network is set up and finished many engineers quickly start the next project without taking time to perform the necessary documentation. There may also be a sense within the networking department that if networks are not documented the networking staff becomes somewhat "irreplaceable." This should not be the case. Well-written documentation saves time and money and should become something you do after any change, large or small, is performed on the network. If someone else can fix your network by reading the documentation you left behind, you have done a good job.

Network Topologies

Before beginning our design, we first review the various *topologies* or architectures available for interconnecting network components. Computers and networks can be connected using a variety of topologies, including bus, ring, star, mesh, point to point, and hybrid. A bus topology uses a single cable that connects all devices in a line. Figure 22-1 depicts a bus topology. Each end of the bus must be terminated. A cable break anywhere on the network takes the entire network offline.

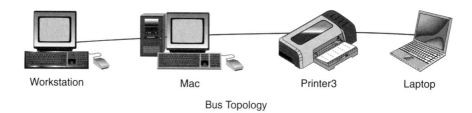

Workstation   Mac   Printer3   Laptop

Bus Topology

FIGURE 22-1   Bus topology.

A ring topology, Figure 22-2, connects all the devices with a central ring. A ring can be thought of as a bus in which both ends are connected to one another. A ring topology does not require termination. A cable break anywhere on the network takes the entire network offline.

In a star topology, Figure 22-3, every node on the network is connected through a central device. In the early days of networking this device was commonly a hub; however, in modern networks the connecting device is a router or a switch. A cabling problem will only affect the connected device, not the entire network.

In a mesh topology, Figure 22-4, every device connects to every other device with two or more cables. The number of cables required depends on the number of devices and can be calculated with the formula:

$$\text{number of connections} = \frac{(\text{number of devices}) \times (\text{number of devices } - 1)}{2}$$

In Figure 22-4 there are five devices; therefore, the number of connections is $\frac{5 \times 4}{2} = 10$.

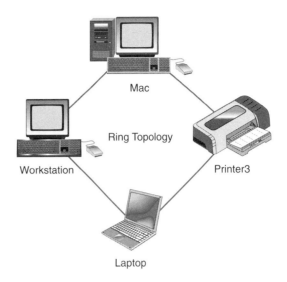

FIGURE 22-2   Ring topology.

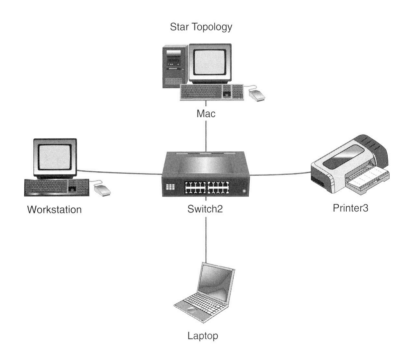

FIGURE 22-3   Star topology.

A point-to-point topology, Figure 22-5, consists of two devices connected together.

Few LANs use the simple physical topologies in their pure form. More often, LANs employ a hybrid of more than one simple physical topology. The star wired bus commonly forms the basis for Ethernet and fast Ethernet networks.

The topology of a network can be considered and, therefore, diagrammed, from either of two perspectives. A *physical topology diagram* represents the basic physical layout of a network's media, nodes, and connectivity devices. It may include information

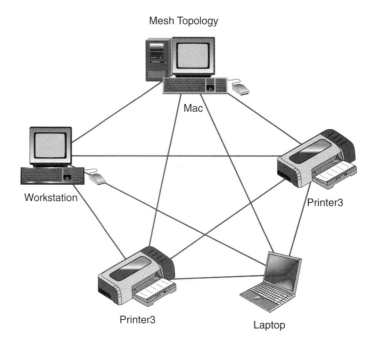

FIGURE 22-4 Mesh topology.

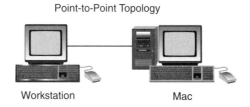

FIGURE 22-5 Point-to-point topology.

about locations within a building, room numbers, placement of outlets within work areas, and the like. A *logical topology diagram* describes how signals travel over a network. The logical diagram is akin to the schematic diagram of an electronic circuit in that it focuses on interconnections and signal flows but does not attempt to present a pictorial representation of individual network pieces. Both types of diagrams have their place, and the following tips will help you professionally diagram your networks from either perspective:

1. Keep it simple: Let the diagram convey a single idea. The more you add to a diagram, the harder it will be to understand.

2. Separate physical and logical ideas: Physical connectivity is important, but try to separate the logical components such as virtual LANs (VLANs) and routing.

3. Do not cross lines: Every time your diagram has crossed lines, it becomes harder to read. Avoiding crossed lines completely may be unavoidable, but try to keep them to a minimum.

4. Orient your straight lines: A diagram with straight lines and similar angles used through out will stand out as a clean, crisp work of art, not some haphazard document that was thrown together at the last minute.

5. **Delineate when you can:** Use colors and shading to differentiate between two groups or locations.

6. **Line up your icons:** If you have a row of icons on your drawing, take the time to line them up along a single axis.

Requirements Document

To build a network we must first gather information from the constituents who want a network built. This information will go into a requirements document. Interviewing stakeholders is the best way to gather this information. Once you have finished gathering the requirements with all the assumptions, write them down and send the finished requirements document to those who need to see it so they can sign off on it. The best time to straighten out exactly what the customer wants is before you purchase the equipment and invest hours of your time. The scope of a project will always change. Scope change is a fact of life with network projects, so do not let this discourage you. At a minimum, the requirements document should address the following areas.

Requirements:

- The number of users the network must support

Assumptions:

- The number of workstations each user will have
- The number of Ethernet interfaces in each workstation
- The required network speed
- The number of telephones and lines per user
- The type of telephone service provisioned (i.e., circuit or IP-based)
- Need to accommodate voice over internet protocol (VoIP) telephony
- Number of users per office or cubicle location
- Number of data and phone outlets per office or cubicle location
- Termination of cable runs

Design

After the requirements are documented and approved, the design phase can begin. At this point, the following questions should be considered:

- How many servers will the network need to support?
- How many printers will be attached to the network, and what will their locations be?
- What applications will be running on the network?
- How will users interact with these applications?
- What type of security is required?
- Is high availability desired and/or affordable?
- How much growth is anticipated?
- Will this network span multiple locations? Multiple floors?

It is recommended that you factor in 30% for growth in network capacity. If the requirements are for 200 ports, configure the switches for 260. This extra capacity will come in handy later. From this information you can begin to think about what type of equipment you will need to purchase. Don't get bogged down in the details at this point because the specifics, including the type and number of products to be procured, will be worked out in consultation with engineering representatives from the vendors to be engaged for the project.

The next thing to create is a sample port assignments spreadsheet like Table 22-1.

TABLE 22-1 Sample Port Assignments Spreadsheet.

PHYSICAL	VLAN/TRUNK/IP	DEVICE	REMOTE INTERFACE
1/1	VLAN 10	Internet Router	F0/0
1/2		Reserved	
1/3			
1/4			
1/5	VLAN 777	Core-2 FWSM Failover	G1/5
1/6	VLAN 777	Core-2 FWSM Failover	G1/6

Next, plan out the IP addressing you will use on your network, such as that in Table 22-2.

TABLE 22-2 Plan for IP Addressing.

NETWORK	MASK	DESCRIPTION
10.1.8.0	255.255.254.0	Internet DMZ
10.1.9.0		
10.1.10.0		Reserved for expansion
10.1.11.0		Reserved for expansion
10.1.12.0	255.255.254.0	VLAN 100
10.1.13.0		

Interconnecting devices requires an understanding of cabling. The two main cable types are copper and optical. The TIA/EIA-568 standard was developed to define telecommunications cabling systems. Current structured network cabling follows the TIA/EIA standard. The standard also defines the pin/pair assignment for eight-conductor twisted pair cabling. There are two varieties of the TIA/EIA-568: A and B. Table 22-3 describes the pin/pair relationship.

A cable with both ends pinned/paired with the same standard, say 568B and 568B, is a straight-through cable. A cable with mixed standards, say 568A and 568B, is a crossover cable. When deciding what type of cable to use, the general rule of thumb is that a crossover cable is used to connect similar equipment, such as PC to PC, and a straight-through cable is used to connect dissimilar equipment, such as a PC to a switch. The one exception to this rule is that a crossover cable is used to connect a PC to a router.

Fiber optic cable provides the following benefits over copper cabling:

- Extremely high throughput
- Very high resistance to noise
- Excellent security

TABLE 22-3   Pin/Pair Relationship.

PIN	568A COLOR	568B COLOR
1	White/Green	White/Orange
2	Green	Orange
3	White/Orange	White/Green
4	Blue	Blue
5	White/Blue	White/Blue
6	Orange	Green
7	White/Brown	White/Brown
8	Brown	Brown

- Long distance without much attenuation
- Industry standard for high-speed networking

Fiber-optic cable comes in two types: multi-mode and single-mode. Multi-mode cabling is typically used over short distances, less than 1000 m. Multi-mode also has a core diameter around 50–100 μm. Multi-mode cabling uses low-cost light-emitting diodes (LEDs) for the light source. Multi-mode cable typically uses an orange jacket. Single-mode cabling is used for longer distances up to 60 km. Single-mode has a core diameter between 8 and 11 μm. Single-mode cable uses an expensive laser for the light source. Single-mode cable typically uses a yellow jacket. Converting from copper to fiber is done with a media converter.

PROCEDURE:

Before beginning this experiment, download and install Network Notepad from http://www.networknotepad.com. Also download and install the Object Library2 and Library3.

Scenario: You have been hired as a networking consultant to build a new network for a company that has just expanded from a single floor in their existing building to the second floor as well. The second floor will be home to the accounting and sales departments. You have met with the vice president of the company, who is overseeing the network expansion project and have determined the following:

- Network speed: All connections must be at 100 Mbps.
- Number of occupants per office: 1.
- Number of phone and data jacks per office: 1 phone (non-IP) port and 2 data ports.
- Need for VoIP capability: There is no foreseeable movement towards VoIP.
- Distances from/to distribution frames: The distance between the main distribution frame (MDF) on the first floor and the intermediate distribution frame (IDF) is approximately 1000 feet.
- Termination of cable runs: All connections on the second floor will homerun to the IDF with the IDF connecting to the MDF and to the rest of the network.
- Maximum distance to IDF: The maximum distance from any office on the second floor to the IDF is less than 90 meters.

In addition, the following design-related questions were answered:

- Number of users (computers): 15
- Number of network printers: 5
- Number of servers: 4 servers
- Growth projection: There is an anticipated growth within the company of at least 15% over the next few years.

You have gathered this information and have verified it with the vice president, and everything has been confirmed to be accurate. Now you will begin to design the network for the second floor knowing that it must interface with the first-floor network. The network administrator on the first floor has provided the following information:

- The existing IP networking will accommodate the additional requirements of the second floor.
- All servers and printers will require a statically assigned IP address in the range of 10.0.0.10–10.0.0.20 with a subnet mask of 255.255.255.0.
- All workstations will receive a dynamically assigned IP address from the servers on the first floor.

1. Create a logical topology diagram using Network Notepad.
    a. Create a separate diagram for the second-floor layout.
    b. Create a separate diagram showing the relationship between the first and second floors.
    c. Show all IP addresses on the diagrams in a tabular format similar to Table 22-2. Also include a port assignments spreadsheet similar to that of Table 22-1.
2. Using real equipment build a model of the network with the following constraints:
    a. Connect a PC (PC1) to a fast Ethernet switch (SW1).
    b. Connect SW1 to a fiber media converter that connects to another media converter on the other end which connects to SW2.
    c. Connect a PC (PC2) to a fast Ethernet switch (SW2).
3. You will have to build all Ethernet cables as needed.
4. Assign the IP address of 10.0.0.1 and 10.0.0.2 to PC1 and PC2.
5. Verify the network works by pinging from 10.0.0.1 to 10.0.0.2.

QUESTIONS:

1. Describe the CSMA/CD process.
2. How would you describe a wireless network in terms of its topology?
3. What is the reason a terminator is used at the ends of a bus network?
4. What are the primary differences between single-mode and multi-mode fiber optic cabling?

LOCAL-AREA NETWORK (LAN) TROUBLESHOOTING

OBJECTIVES:

1. To become familiar with what troubleshooting is.
2. To understand the Open Systems Interconnection (OSI) model and relate this information to troubleshooting.
3. To understand top-down troubleshooting.
4. To understand bottom-up troubleshooting.
5. To understand divide-and-conquer troubleshooting.

REFERENCE:

Refer to Sections 11-3 through 11-5 in the text.
http://www.flukenetworks.com/expertise/learn-about/Troubleshooting-LANs

EQUIPMENT:

Two or more PCs connected to a network

INTRODUCTION:

Local-area networks (LAN) are an integral part of nearly every business today. The most common LANs use Ethernet, a data link layer protocol, and Internet Protocol (IP), a network layer protocol.

A local-area network has many parts to it. These parts may include printers, monitors, PCs, IP phones, servers, storage devices, networking equipment, security software, network applications, office applications, and many more. Devices connect to the network using twisted-pair copper cabling, fiber-optic cabling, or wireless access points (WAPs).

Troubleshooting LANs is normally in the job description of the network support staff, which is made up of communication engineers and technicians. Common problems include user connection issues and slow network performance.

There are frequently three root causes for LAN problems:

1. Physical layer: copper, fiber, or wireless
2. Network layer: Ethernet and IP
3. Switches and virtual local-area networks (VLANs)

Outright failures are easily detected; when a router fails completely, there are very obvious symptoms that affect many people. When a switchport starts to fail by mangling some packets, the problem can be quite a bit harder to diagnose.

Troubleshooting is both an art and a science. Troubleshooting can be represented with a very basic equation: problem + diagnosis = solution. One of the best ways to become an expert troubleshooter is to combine knowledge of the scientific method and the OSI model.

There are six steps in the scientific method:

1. Gather information.

2. State the problem.

3. Form a hypothesis.

4. Test the hypothesis.

5. Observe results and draw conclusions.

6. Repeat as necessary.

The steps for troubleshooting include:

1. Identify the problem and scope.

2. Establish probable cause.

3. Test the theory.

4. Establish a plan of action.

5. Implement the solution.

6. Verify and implement preventative measures.

7. Perform root-cause analysis.

8. Document.

There are two types of troubleshooters: those who take a stab in the dark to solve the problem and those who take a systematic approach. The stab-in-the-dark approach usually involves little knowledge of the technology involved and is mostly random in nature. A systematic approach is step-by-step and requires an in depth knowledge of the technology. This lab will utilize the systematic approach to troubleshooting.

The first step to troubleshooting is to identify the problem and the scope. Normally, end users are very good at letting the network administrator know when there is a problem. You must first define the scope of the problem. When defining the scope, you need to identify what is and what is not working.

Once a scope is identified, the next phase is to establish a probable cause. This includes information gathering, analysis, process of elimination, and proposing a hypothesis. Observation and user interviewing are the most common ways to establish a probable cause. A user's report of the problem may not contain enough information, and a direct observation of the problem should be sought. A repeatable problem that can be recreated can be observed more easily than a transient or intermittent problem. This phase of troubleshooting is normally the most involved and takes the most amount of time.

Once you have a solid hypothesis, test it out. Confirm or deny the theory. It is extremely important to make only a single change at a time. Working with single variable changes to the equation is more instructive than changing multiple things all at once. If the variable that you change does not solve the problem, undo the change. Now that we have honed in on the problem, we can plan our actions to perform the fix. The least invasive solutions should be attempted first. You do not want to make things worse by being overly aggressive in your plan. Establish a plan of action; do not just dive into the situation and start trying to repair the problem. Being methodical may take longer, but it will also prevent making costly mistakes. Once you have planned how to fix the problem, fix it. There are several ways to implement the solution: *top down, bottom up, divide and conquer, follow the path, spot the differences,* and *move the problem.*

A *top-down approach* involves the OSI model, starting with the application layer and working your way down to the physical layer, as shown in Table 23-1. In a top-down approach, access to the client is needed at the beginning of the implementation of the solution. This approach is the best to use if multiple users initiate several help desk calls.

TABLE 23-1   The Top-Down Approach Using the OSI Model.

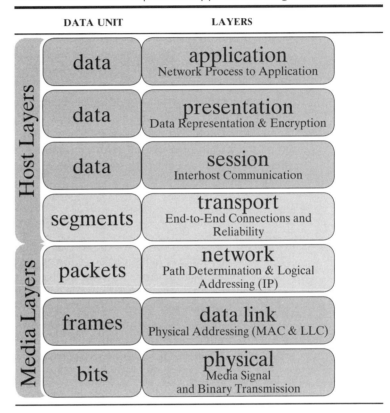

DATA UNIT	LAYERS
Host Layers	
data	**application** Network Process to Application
data	**presentation** Data Representation & Encryption
data	**session** Interhost Communication
segments	**transport** End-to-End Connections and Reliability
Media Layers	
packets	**network** Path Determination & Logical Addressing (IP)
frames	**data link** Physical Addressing (MAC & LLC)
bits	**physical** Media Signal and Binary Transmission

The *bottom-up approach* to troubleshooting starts with the physical components and works its way up the OSI model. Bottom-up troubleshooting is an effective and efficient approach for situations when the problem is suspected to be physical. Most networking problems reside at the lower levels, so implementing the bottom up approach often results in effective and perhaps faster results. The downside to the bottom up approach is that it requires you to check every device, interface, and so on. In other words, regardless of the nature of the problem, the bottom-up approach starts with an exhaustive check of all the elements of each layer, starting with the physical layer.

The *divide-and-conquer approach* does not necessarily start its investigation at any particular OSI layer. When you apply the divide-and-conquer approach, you select a layer and test its health; based on the observed results, you might go in either direction (up or down) from the starting layer. If a layer is in good working condition, you inspect the layer above it. If a layer is not in good working condition, you inspect the layer below it. The layer you start at is based on your experience level and the symptoms you have gathered about the problem. If you can verify that a layer is working well, it is safe to assume the layers below it are functioning properly as well. If a layer is not functioning at all, or is working intermittently or erroneously, you must inspect the layer below it. If the layer below it is working properly you can safely assume the problem is the current layer.

The *follow-the-path* troubleshooting approach involves following a packet or frame as it traverses the network. The *spot-the-difference* approach is used when you can compare a working system to a non-working system. It is possible to "fix" a problem without knowing the root cause. A *move-the-problem* solution isolates the problem by swapping out components. If the problem moves with the components, the problem is in that component. This is not an effective method if the problem spans more than one component.

So which approach should you use? Selecting the most effective method involves the following: determining the scope of the problem, applying your experience, and analyzing the symptoms. As we stated earlier, troubleshooting is both an art and a science. Troubleshooting is an iterative process. Some of the previous steps (follow the path, spot the difference, and move the problem) may solve the problem without identifying what caused the problem in the first place. Once the root cause is identified, you can take preventive actions. Implementation of preventive measures serves to fix and also prevent the recurrence of a problem. This leads to continual improvement in your overall information technology processes. It is recommended that you fully document your troubleshooting experiences to build a knowledge base for your environment.

PROCEDURE:

1. Working in pairs or groups, introduce into your computer(s) the five faults that follow. Work with one fault at a time. Once your group has introduced a fault, invite another user or group over to your computer and have them troubleshoot the problem. Once you are done with each fault, be sure to undo it. Alternatively, each fault may be introduced into a separate computer; each user or group may then come up to the computer and troubleshoot the problem. Your instructor will advise as to the approach to be used in your laboratory.

2. For each of the five faults, list the steps you took to diagnose and repair the problem. Include the following information for each fault:

   a. Identify the problem and scope.

   b. Establish probable cause.

   c. Test the theory.

   d. Establish a plan of action.

   e. Implement the solution.

   f. Verify and implement preventive measures.

   g. Perform root-cause analysis.

   h. Document.

Fault 1: Change the IP address of the PC to a IP address that is not on the same network, such as 192.168.0.1 with a subnet mask of 255.255.255.0.

Problems encountered: Cannot connect to the internet or other network resources.

Fault 2: Change the network adapter properties to 10 Mbps, half duplex.

Problems encountered: Your network connection changes from on to off repeatedly.

Fault 3: Change the default gateway from its current value to 10.254.254.254.

Problems encountered: You have full connectivity and access to local network resources but not to anything outside of your network, such as the internet.

Fault 4: Connect your PC to the network with a crossover cable instead of a patch cable.

Problems encountered: You have no connectivity to any network resources.

Fault 5: Edit the hosts file located at c:\windows\system32\etc\ by adding the line:

127.0.0.1   http://www.google.com

Problem encountered: You can access all websites with the exception of http://www.google.com

QUESTIONS:

1. If a user opens a trouble ticket after a new software application has been installed on their system, what approach would make the most sense to apply first? Why?

2. What approach is normally the fastest when troubleshooting networking problems? Why?

3. If users report that they cannot access a particular web page, but can access all other web pages problem free, what approach should be used? Why?

4. Briefly describe the seven layers of the OSI model.

BINARY AND IP ADDRESSING

OBJECTIVES:

1. To explore number systems including decimal, binary, octal, and hexadecimal.
2. To convert between various number systems.
3. To understand the core functionality of IP addressing and subnetting in networking.
4. To become familiar with *classful* and *classless* IP addresses.

REFERENCE:

Refer to Section 11-6 in the text.

EQUIPMENT:

None

COMPONENTS:

None

INTRODUCTION:

Converting between various number systems is an essential task all communications professionals must be able to perform. Most calculations you have performed have been based on the decimal number system, a positional number system in which each integer in a decimal number has a value associated with it. For example, 843 in decimal is 8 hundreds + 4 tens + 3 ones. From right to left, the value of each position is 10^0, 10^1, 10^2, 10^3, and so on. (Note that in this context, exponents are always 0 or positive.) The term decimal refers to the number of possible values, which in this case is 10. These values are 0–9.

Computers and all networking equipment understand binary numbers only. Binary numbering systems have only two possible values: 0 or 1. In terms of voltage, 1 corresponds to the presence of a voltage level and 0 corresponds to a lack of voltage. The binary system is also positional in that each position from right to left corresponds to the base raised to a value. The positions 2^7, 2^6, 2^5, 2^4, 2^3, 2^2, 2^1, and 2^0 represent the weighted values of an eight-digit binary number, also known as a byte, as shown in Table 24-1.

2^7	2^6	2^5	2^4	2^3	2^2	2^1	2^0
128	64	32	16	8	4	2	1

Other number systems use weighted values as well. But each system uses its own base numeral and has its own number of positional values. For example, in an octal system, the base numeral is 8 and there are 8 positional values from 0 to 7. In the hexadecimal system, the base numeral is 16 and there are 16 positional values, consisting of the digits 0 through 9 and the letters A through F, which represent the values 10 through 15. Once this positional weighting is understood, conversion between different number systems becomes a straightforward endeavor.

As mentioned in Section 11-6 in the text, networked computers and equipment are assigned IP addresses. On any network each connection or node must have a unique IP address. If you have ever worked with a networked computer you may have seen these IP addresses. If you open a command prompt on your Windows computer and type the "Ipconfig" command, you will see a reply with an IP address such as 192.168.2.100. This dotted decimal number represents four different decimal numbers, with each one representing an 8-bit number or octet.

But how does your computer get assigned this IP address? There are two ways IP addresses are assigned: statically and dynamically. Static IP addresses are manually assigned and are not practical on larger networks, say any network with more than ten nodes. Dynamically assigned IP addresses require a dynamic host configuration protocol (DHCP) service to be available on the network. DHCP can be performed by a router or by a service on a server. DHCP is recommended on networks with more than ten nodes. Certain devices should always be statically assigned and others should be dynamically assigned. Servers, routers, switches, network printers, and wireless access points should have a static IP address, while workstations should have dynamically assigned addresses.

In this experiment, we will work with IPv4 addressing. IPv4 uses a 32-bit address space. All told, there are around 4.3 billion IP addresses. The newest iteration of IP addressing is known as IPv6, which uses 128 bits. IPv6 will not be covered in this experiment.

IP addresses are classified based on the bits in their first octet. First, let's take a look at the major network classes in Table 24-2.

TABLE 24-2 Major Network Classes.

CLASS	START	END	NETWORKS	HOSTS PER NETWORK
A	0.0.0.0	127.255.255.255	128	16,777,214
B	128.0.0.0	191.255.255.255	16,384	65,534
C	192.0.0.0	223.255.255.255	2,097,152	254
D	224.0.0.0	239.255.255.255	268,435,456	0
E	240.0.0.0	255.255.255.255	Undefined	Undefined

Class A addresses always have the first bit of their IP address set to "0". Class B addresses have the first bit set to "1" and the second bit set to "0". Class C addresses have the first two bits set to "1" and the third bit set to "0". Class D addresses are used for multicasting applications and have the first three bits set to "1" and their fourth bit set to "0". Class E addresses are defined as experimental and are reserved for future testing purposes.

At one time, if an organization requested a class A address space from their service provider they would potentially be using over 16 million IP addresses. If the organization really only needed 100,000 IP addresses, then over 15.9 million addresses would be unused or wasted. A class B address space took up over 65,000 addresses. In a relatively short amount of time the organization that handles the distribution of all IPv4 addresses, the Internet Assigned Numbers Authority (IANA), ran out of available addresses. The "one size fits all" classful address provisioning was not efficient in rightsizing IP addresses.

This problem has been resolved through the use of variable length subnet masks. A subnetwork, or subnet, is the division of a network into two or more networks. This process is known as subnetting. Subnet masks are shown by network class in Table 24-3.

TABLE 24-3 Subnet Masks by Network Class.

CLASS	SUBNET MASK (DECIMAL)	SUBNET MASK (SLASH)
A	255.0.0.0	/8
B	255.255.0.0	/16
C	255.255.255.0	/24

The slash notation refers to the number of 1s in the subnet mask. The number of 1s references the number of networks; the number of 0s references the number of hosts per network. In a class A network there are eight 1s and 24 0s; therefore, the number of networks is 2^8 or 128 and the number of hosts is 2^{24} or 16,77,214. If an IP address scheme calls for a design that can accommodate up to 500 hosts a single class C address will not work, as this can only handle up to 254 hosts. In this case the subnet mask should be 255.255.254.0. In this example the number of 0s would be 9 so the number of hosts would be 2^9 or 512.

PROCEDURE:

Part I: Number Conversions

1. To begin, let's convert a *binary number to decimal.* For this example, let's convert 10100011, an eight-digit binary number.

   Step 1: Write down the weighted values from Table 24-1.

128	64	32	16	8	4	2	1

   Step 2: Write down the binary number under the each weighted value.

128	64	32	16	8	4	2	1
1	0	1	0	0	0	1	1

   Step 3: Now add up the weighted values where there are 1's.

   $$128 + 32 + 2 + 1 = 163; \text{ therefore, } 10100011_2 = 163_{10}.$$

2. To convert from *decimal to binary* we start the same way, by writing down our weighted values. For this example let's convert 248_{10} to binary.

Step 1: Write out the binary system.

128	64	32	16	8	4	2	1

Step 2: Is 248 greater than or equal to 128? If the answer is yes, write a 1 under 128.

Is 248 >= 128? Yes. Write a 1 under 128.

128	64	32	16	8	4	2	1
1							

Step 3: Subtract 128 from 248.

$$248 - 128 = 120.$$

Now ask yourself: Is 120 >= 64? Yes. Write a 1 under 64.

128	64	32	16	8	4	2	1
1	1						

Step 4: Subtract 64 from 120.

$$120 - 64 = 56.$$

Is 56 >= 32? Yes, Write a 1 under 32.

128	64	32	16	8	4	2	1
1	1	1					

Step 5: Subtract 32 from 56.

$$56 - 32 = 24.$$

Is 24 >= 16? Yes. Write a 1 under 16.

128	64	32	16	8	4	2	1
1	1	1	1				

Step 6: Subtract 16 from 24.

$$24 - 16 = 8.$$

Is 8 >= 8? Yes. Write a 1 under 8.

128	64	32	16	8	4	2	1
1	1	1	1	1			

Step 7: Subtract 8 from 8.

$$8 - 8 = 0.$$

The rest of the values in the table are 0s.

128	64	32	16	8	4	2	1
1	1	1	1	1	0	0	0

We read the conversion from decimal to binary from the table: $248_{10} = 11111000_2$.

3. Hexadecimal is a number system with a base of 16. The hexadecimal characters are shown in Table 24-4.

TABLE 24-4 Characters of the Hexadecimal System.

VALUE	DECIMAL	BINARY
0	0	0000
1	1	0001
2	2	0010
3	3	0011
4	4	0100
5	5	0101
6	6	0110
7	7	0111
8	8	1000
9	9	1001
A	10	1010
B	11	1011
C	12	1100
D	13	1101
E	14	1110
F	15	1111

4. To convert a *binary number to hexadecimal,* divide it into groups of four digits starting with the rightmost digit. If the number of digits is not a multiple of 4, prefix the number with 0's so that each group contains 4 digits.

For example, convert 10110101_2 to hexadecimal.

Separate into groups of 4 digits:	1011	0101
Convert each group to hex digit:	B	5

Therefore: $10110101_2 = B5_{16}$.

Note: Hexadecimal is commonly shown with the $0 \times$ prefix such as $0 \times B5$.

5. Converting from *hexadecimal to binary* is very similar in that you take the hexadecimal number and write its four-digit binary equivalent and put them all together.

For example, convert $6B8C_{16}$ to binary.

Separate each digit:	6	B	8	C
Convert each group to binary digit:	0110	1011	1000	1100

Therefore $6B8C_{16}$ or $0 \times 6B8C = 0110101110001100_2$

6. Converting from *binary to octal* is very similar, except we group our binary values in groups of three digits.

For example, convert 111110011001_2 to octal.

Separate into groups of 3 digits:	111	110	011	001
Convert each group to decimal:	7	6	3	1

Therefore: $111110011001_2 = 7631_8$

7. Converting from *octal to binary*.

For example, convert 536_8 to binary.

Separate each digit:	5	3	6
Convert to 3 digit binary value:	101	011	110

Therefore $536_8 = 101011110_2$.

All other conversions can be done by converting to binary first and then to the desired number system.

Part II: IP Addressing

Now let's work on a few IP-addressing problems. As discussed in Section 11-6 of the text, an IP address consists of two parts: the network ID and the host ID. The network ID, also known as a network address, identifies a single network segment within a larger TCP/IP network. All of the systems that attach and share access to the same network have a common network ID within their full IP address. This ID is also used to uniquely identify each network within the larger network. The host ID, also known as the host address, identifies a TCP/IP node (a workstation, server, router, or any other TCP/IP device) within each network. The host ID for each device identifies a single system uniquely within its own network. Hosts and routers use Boolean math to determine the network ID and the host ID.

The truth table for the Boolean AND function is shown in Table 24-5.

TABLE 24-5 Truth Table for the Boolean AND Function.

BIT 1	BIT 2	RESULT
0	0	0
0	1	0
1	0	0
1	1	1

8. For example, what is the network ID of the IP node 129.56.189.41 with a subnet mask of 255.255.240.0?

Step 1: Convert all dotted decimals into binary.

129.56.189.41 =	10000001	00111000	10111101	00101001
255.255.240.0 =	11111111	11111111	11110000	00000000

Step 2: AND the binary representations together as follows:

	10000001	00111000	10111101	00101001
AND	11111111	11111111	11110000	00000000
=	10000001	00111000	10110000	00000000
=	129.56.176.0			

Thus, 129.56.176.0 is the network ID of the IP node 129.56.189.41 with a subnet mask of 255.255.240.0.

Recall that in a subnet mask, the 1s represent the network and the 0s represent the hosts. Subnet masks can be written in slash notation, which shows the number of 1s in the subnet mask; therefore, 255.255.255.128 would be /25. If you have a /25 subnet this means there are 25 1s and 7 0s. You can determine the number of hosts by using the following formula:

$$2^n - 2 = \text{available hosts where } n \text{ is the number of 0's.}$$

We subtract 2 because the first address is the network ID and the last address is the broadcast ID. If we have a /24 subnet mask we could have $2^8 - 2 = 254$ hosts. For instance, if the network ID is 192.168.2.0, the broadcast address is 192.168.2.255, the first usable IP address is 192.168.2.1, and the last usable IP address is 192.168.2.254.

If you are given a number of hosts and are asked to come up with the best subnet mask to use to prevent wasting IP addresses, the best way to tackle this problem is to look at the host part of the subnet mask and work backwards.

QUESTIONS:

1. Convert the following numbers.

   (a) 10110011_2 to decimal

   (b) $A95_{16}$ to binary

   (c) 7564_8 to hexadecimal

   (d) $240.187.35.17_{10}$ to binary

   (e) 01111010_2 to hexadecimal

   (f) 65254_{10} to octal

2. On a network with an IP address of 140.133.28.72 and a subnet mask of 255.248.0.0 what is the network ID?

3. As a networking consultant, you have been asked to help expand a client's TCP/IP network. The network administrator tells you that the network ID is subnetted as 185.27.54.0/26. On this network, how many bits of each IP address are devoted to host information?

4. You are tasked with designing a network that can support forty-five computers, ten servers, four printers, and three WAPs. Assume that your IT infrastructure will increase by 10% over the next five years. What is the best network mask to use for this scenario?

STANDING-WAVE MEASUREMENTS OF A DELAY LINE

OBJECTIVES:

1. To take data for plots of the voltage standing-wave patterns of a simulated transmission line terminated with open, shorted, and matched loads.

2. To use data and a Smith chart to determine the value of an unknown impedance.

3. To use ac circuit theory to determine the value of an unknown impedance.

REFERENCE:

Refer to Section 12-6 in the text.

EQUIPMENT:

Dual-trace oscilloscope

Function generator

Volt-ohmmeter

Pulse generator

Impedance bridge

COMPONENTS:

Resistors: 10 Ω, 22 Ω, 180 Ω, 270 Ω, and other values needed to obtain a proper match between generator and transmission line (explained in the procedure)

Potentiometer: 500 Ω

Capacitors: 0.001 μF, 0.01 μF (20), 0.1 μF, 0.22 μF

Inductors: 1 mH (20)

INTRODUCTION:

If an impedance is given, the voltage standing-wave ratio (VSWR) of a system can be obtained from a specialized graph called the Smith chart. The radius from the center of the Smith chart to the impedance (written and plotted in normalized form) can be used to determine the VSWR on the standing-wave ratio axis of the Smith chart. Conversely, if the VSWR is known, a circle can be drawn on the Smith chart that passes through the coordinates of all possible normalized impedances. Refer to Figure 25-1.

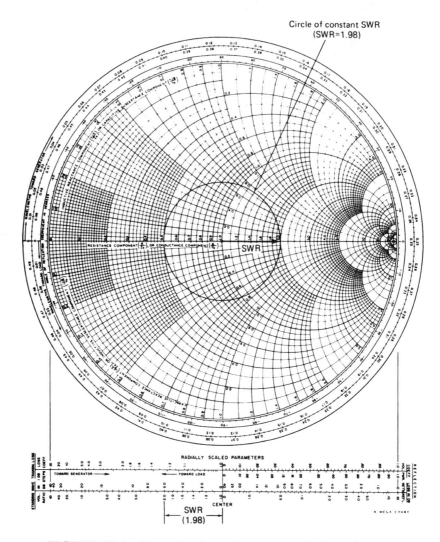

FIGURE 25-1 Example of standing-wave ratio (SWR) circle.

The problem then remains of determining which point on the circle represents the actual impedance. The problem can be solved by comparing the location on the transmission line of a voltage minimum referred to as a null caused by the unknown load with the location of a voltage null caused by a short.

Any mismatch on the transmission line will cause a standing wave to result. In other words, sinusoidal voltages measured at various points on the line will have different amplitudes. The change in amplitude with respect to location will exhibit a regular pattern. This pattern, which has the same shape as that of a full-wave rectifier's waveform, is referred to as the standing-wave pattern, as shown in Figure 25-2. Keep in mind, however, that this is a plot of the magnitude of the sine wave versus distance away from the load, not versus time. Thus do not think of this pattern as a waveform itself.

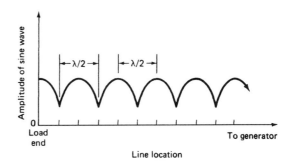

FIGURE 25-2 Standing wave when $R_L > Z_o$.

A short circuit will cause a voltage null to occur at the load end of the line and additional nulls every $\lambda/2$ distance away from the load toward the generator. "λ" represents the wavelength of the signal being transmitted down the transmission line. It can be computed by dividing the velocity of the signal passing through the line by its frequency ($v = f\lambda$). Any purely resistive load with an ohmic value less than the characteristic impedance of the line, Z_o, will cause voltage nulls and peaks to occur at the same locations as those observed when the load is a short, but with different amplitudes. Refer to Figure 25-3.

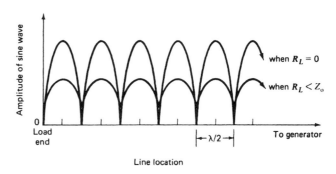

FIGURE 25-3 Voltage standing wave of a short when $R_L < Z_o$.

An open circuit will cause a voltage peak at the load end of the line, a null at a location $\lambda/4$ down the line, and additional nulls at locations every $\lambda/2$ after that. Any purely resistive load with an ohmic value greater than Z_o will cause voltage nulls and peaks to occur at the same locations as those observed when the load is an open circuit, but with different amplitudes. Refer to Figure 25-4.

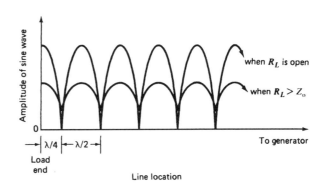

FIGURE 25-4 Voltage standing wave when $R_L > Z_o$.

Looking at the Smith chart, you can see that any resistive load quantity that is less than Z_o would be plotted on the real axis between zero and 1.0. Also, you can see that any load of this type would have the same reference location on the wavelength scale of the Smith chart, namely, 0λ. On the other hand, any resistive load quantity that is greater than Z_o would be plotted on the real axis between infinity and 1.0. The wavelength location in this case would be 0.25λ.

Referring back to Figures 25-3 and 25-4, you can observe that there is a direct correlation between the location of the voltage standing-wave nulls on the transmission line and the wavelength reference locations on the Smith chart. Any resistive impedance value that is less than Z_o gives a voltage null at location zero on the transmission line and a reference location 0λ on the Smith chart. Any resistive impedance value that is greater than Z_o yields a voltage null at a location $\lambda/4$ away from point zero on the line and is also plotted at the reference location 0.25λ on the Smith chart. Thus, it can be deduced that any complex impedance, either inductive or capacitive, will cause voltage nulls to fall at line locations other than 0λ or $\lambda/4$ with respect to nulls caused by a short. Refer to Figure 25-5. The location of these voltage nulls and the VSWR can be used to determine the actual value of an unknown load impedance. The following steps summarize the proper procedure to be used.

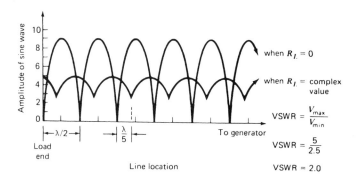

FIGURE 25-5 Voltage standing wave of a short and complex load.

1. From the maximum and minimum voltage amplitude data, calculate the VSWR. (VSWR = 2.0 in example)

2. Draw the VSWR circle on the Smith chart using the calculated value of the VSWR to set the radius of the circle.

3. Find the location of a voltage null using the standing wave pattern for the transmission line terminated with the unknown load. Determine the distance in wavelengths toward the load (clockwise) to the first encountered voltage null on the standing wave pattern for the transmission line terminated with a short circuit. (distance = $\lambda/5$ in example)

4. Draw a straight line on the Smith chart from the center of the chart to the same location on the "wavelengths toward the load" scale that represents the distance determined in step 3.

5. Read the normalized impedance of the load at the intersection of the straight line and the VSWR circle. ($Z_{Ln} = 1.55 - j0.65$ in example)

6. Calculate the unknown load impedance by multiplying the normalized value obtained in step 5 by the characteristic impedance of the line, Z_o. ($Z_L = 77.5 - j32.5\ \Omega$ in example)

Refer to Figure 25-6 for the solution to the example problem given in Figure 25-5.

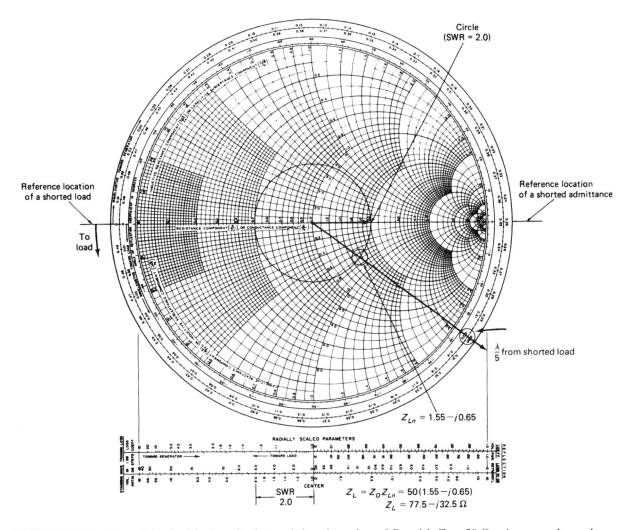

FIGURE 25-6 Use of the Smith chart in determining the value of R_L with $Z_o = 50\ \Omega$ using complex voltage.

The distance between voltage nulls also provides a convenient way to calibrate the transmission line in units. The distance between voltage nulls must always be $\lambda/2$. Matching of a variable resistive load can also be achieved by using the short circuit data if the transmission line has negligible losses. The location of a short-circuit peak and null voltage can be monitored simultaneously. The load can then be adjusted until the voltage at the peak location is equal to the voltage at the null location. Under these conditions, the voltage everywhere on the line must be equal and the line must be matched. The sinusoidal voltage observed at the peak and null locations is $\lambda/4$ or $90°$ out of phase. Line loss will introduce some error into these impedance matching and measuring techniques. The effect of line loss can be observed in this experiment, but it will have minimal effect on the impedance measurement. Refer to Figure 25-7 to observe the effects of line loss on the standing-wave pattern.

PROCEDURE:

1. Using the given values of the components within the complex load of Figure 25-8 and the given frequency of 10 kHz, calculate the theoretical value for the complex impedance and express in rectangular coordinates.

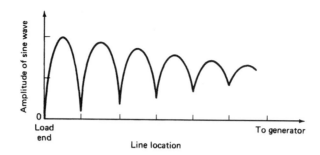

FIGURE 25-7 Voltage standing wave of a short on a line containing a high resistive loss.

2. Construct the test configuration as shown in Figure 25-9. Short the load end of the transmission line. Apply a 50-μs pulse width, 2.0-ms period pulse to the input of the transmission line. Adjust the matching network as necessary to obtain a match at the generator end of the transmission line. This is done by adjusting resistors R_1, R_2, and R_3 to eliminate any re-reflections from occurring at the generator end of the line. Measure and record the values of R needed for a matched impedance condition. The values of R_1, R_2, and R_3 given in Figure 28-9 are for a 50-Ω generator. If your generator has other than a 50-Ω impedance, other values of R_1, R_2, and R_3 will have to be used. In all cases, the pulse generator and function generator must have the same internal impedance for this procedure to be useful.

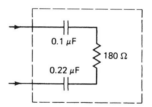

FIGURE 25-8 Complex load.

3. Disconnect the pulse generator and connect the function generator in its place. Apply a 10-kHz sinusoidal voltage of convenient amplitude to the circuit of Figure 25-9. Check the frequency with a frequency counter. Connect the potentiometer to the load end of the line and match it to the line. Do this by following the procedure given at the end of the theory section of this experiment. With the load matched, measure each of the twenty-one node voltages and record all measured values in Table 25-1. Also, disconnect the potentiometer and measure its matched resistance value. It should be close to 300 Ω.

4. Open the load end of the transmission line. Measure the amplitudes of each of the sinusoidal waveforms observed at the twenty-one nodes on the line and record in Table 25-1.

5. Short the load end of the line. Again, measure each of the twenty-one node voltages and record in Table 25-1.

6. Connect the complex load (Figure 25-8) to the transmission line. Again, measure each of the twenty-one node voltages and record values in Table 25-1.

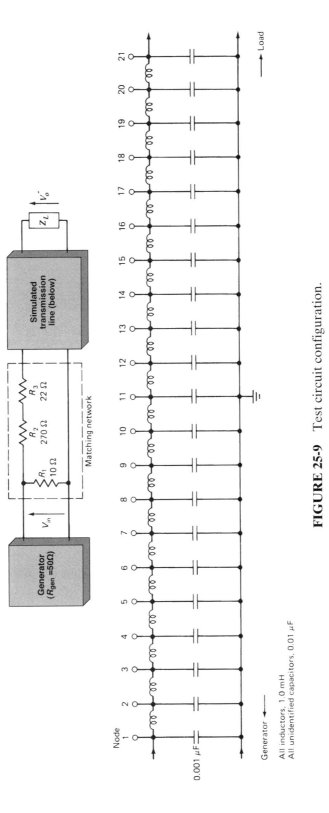

FIGURE 25-9 Test circuit configuration.

TABLE 25-1   Measured Node Voltages with Open, Short, Matched, and Unknown Loads.

| NODE | VOLTAGE (V_{P-P}) | | | |
	R_L (OPEN)	R_L (SHORT)	R_L (MATCHED)	R_L (UNKNOWN)
1				
2				
3				
4				
5				
6				
7				
8				
9				
10				
11				
12				
13				
14				
15				
16				
17				
18				
19				
20				
21				

7. Measure the exact value of each component in the complex load using an impedance bridge if one is available.

8. Make sure that you have gathered sufficient data to complete each of the items given in the report section of this experiment.

QUESTIONS:

1. On one sheet of graph paper, plot the open, shorted, and matched-load-impedance standing-wave patterns. Do this by plotting voltage amplitude versus line location away from the load. Calibrate the location axis in nodes and in quarter-wavelength sections. Draw all three patterns on the same set of axes. Refer to Figures 25-2, 25-3, and 25-4.

2. On a second sheet of graph paper, plot the shorted and complex impedance standing-wave patterns. Use the same plotting technique as in question 1. Refer to Figure 25-5.

3. Using the plots drawn in the question above and the Smith chart, determine the value of the complex impedance. Express this value in rectangular coordinates. The characteristic impedance of the simulated transmission line is approximately 300 Ω, so use this value in your calculations.

4. Repeat question 3, but this time use the measured Z_o value determined in step 3 of the test procedure.

5. Describe fully all results of this experiment, with particular attention to relevant comparisons. Be numerically specific. In particular, compare your Smith chart values of the unknown load impedance to the calculated value based on circuit analysis.

6. What factors in this experiment may have led to discrepancies in the resulting values of the complex load impedance? Which value do you feel is the most accurate? Why?

USING CAPACITORS FOR IMPEDANCE MATCHING

OBJECTIVE:

To use a Smith chart for determining the size and location of series and shunt impedance-matching capacitors.

REFERENCE:

Refer to Section 12-8 and 12-10 in the text.

EQUIPMENT:

None

COMPONENTS:

Smith charts (11″×17″)
Compass
Ruler

INTRODUCTION:

A Smith chart is input impedance mapped on to the plane of reflection coefficient. Moving clockwise along a circle (centered in the middle of the Smith chart) shows how the input impedance is "transformed" to different values corresponding to moving towards the RF generator. Refer to Figure 12-28 in the text to review the construction of a Smith chart.

PROCEDURE:

Part I: Circuit Description

1. A coaxial transmission line has the following parameters: $Z_o = 50\ \Omega$, $E_r = 1$, and frequency = 1 GHz. Compute wavelength in centimeters. $C = 3 \times 10^{10}$ cm per second.

_____ (a)

Part II: Measuring Z_{in} along the Transmission Line

2. On the Smith chart, mark the location for $Z_L = 25\ \Omega$.

3. Using a compass, draw a circle, centered at $Z_{in} = 50\ \Omega$ and passing through the point noted in step 2.

4. What is the VSWR for all points on this circle?

_____(b)

5. Compute the number of wavelengths for a distance of 3 cm.

_____(c)

6. From the point identified in step 2, move along the circle (towards the generator) by 3 cm. Using a ruler measure $Z_{in}/Z_o =$ _____ $+$ j _____ (d)

7. Calculate $Z_{in} =$ _____ $+$ j _____ (e)

Part III: Impedance Matching with a Series Capacitor

8. Find the number of wavelengths from the point in step 2 required to make $R_e\ (Z_{in}) = 50\ \Omega$. To do so, move clockwise along the VSWR circle to the point where the VSWR circle intersects the resistance circle corresponding to a normalized resistance value of 1.0. Then, extend a straight line from the origin through the point where the circles intersect and out to the scale labeled "wavelengths toward generator" along the circumference of the chart. Use that scale to read the distance in wavelengths. _____(f)

9. Compute the corresponding physical distance by multiplying the number of wavelengths computed in (f) above by the distance per wavelength you determined in Part I (a). _____(g)

10. What is the magnitude and phase of the reflection coefficient at this point? The phase of the reflection coefficient is read from the scale labeled "angle of reflection coefficient in degrees" along the outer edge of the chart. Next, measure the radius of the VSWR circle using a compass or ruler. Then, move the measuring instrument to the "radially scaled parameters" scale below the chart. Find the scale labeled "rfl coeff, E or I." The magnitude of reflection coefficient is equal to the distance from the center of that scale to a point left of center and equal to the radius of the VSWR circle. _____(h)

11. What is $I_m\ (Z_{in})$? The imaginary component of impedance can be determined by reading the reactance arc that also intersects the point determined by the intersection of VSWR and resistance circles identified in step 1 of this section. To determine I_m follow your finger up the arc to the scale along the outer edge of the chart labeled "Inductive reactance component." _____(i)

12. The value determined as (i) in the previous step is a value of inductive reactance. The amount of capacitive reactance needed is equal and opposite in value to that of inductive reactance. Compute the amount of series capacitive reactance required to make the reflection coefficient equal to zero. _____(j)

13. Compute the corresponding capacitance for step 12. The appropriate capacitance can be determined by rearranging the formula for capacitive reactance and using the design frequency of 1 GHz and the reactance recorded as (j). _____(k)

Part IV: Impedance Matching with a Shunt Capacitor

14. Start at a point defined in $Z_L = 15 + j15 \; \Omega$.

15. Draw a circle through this point.

16. In order to convert from Z_{in} / Z_o to Y_{in} / Y_o: Draw a line through this point and the center of the Smith chart and extend it through the other side of the circle. Y_{in} / Y_o is located at the other intersection.

17. Note the value located along the scale "wavelength towards generator" and record. _____(l)

18. Follow the circle further towards the generator (i.e., continue clockwise) until R_e ($Y_{in}/Y_{out} = 1 - j0$). This point is represented where the VSWR and unity resistance circles intersect. Extend a straight line from the origin through the point where the circles meet and to the outer edge of the chart.

19. From the "wavelengths toward generator" scale, record the wavelength towards the generator and record. _____(m)

20. What is the physical distance between step 17 and step 19? Determine the physical distance first by subtracting (m) from (l) to determine the wavelength distance and then by multiplying the result by the distance in centimeters for one wavelength determined in (a). _____(n)

21. What is I_m (Y_{in} / Y_o)? This value is determined in a manner similar to the determination of the imaginary component of impedance in step 11: Read the arc that also intersects the point determined by the intersection of VSWR and resistance circles. Follow your finger down the arc to the scale along the outer edge of the chart labeled "Capacitive reactance component" and record. _____(o)

22. Compute Y_{in} by denormalizing the values found in step 21. Denormalization of admittances requires that each normalized value be divided by the characteristic impedance (50 Ω in this case) rather than multiplied, as in previous steps. The normalized real value for admittance is 1.0, and the normalized imaginary value is the value you recorded in step 21.

_____ –j _____(p)

23. Compute the value of shunt capacitance required to make the reflection coefficient equal to zero. This is a three-step process mirroring steps 11 through 13. First, take the imaginary (j) value recorded in line (p) (step 22) and denormalize it by dividing by the system impedance 50 Ω. The result is the capacitive reactance. Again reorganize the formula for capacitive reactance as you did in step 13 to compute the value of shunt capacitance to make the reflection coefficient equal to zero. _____(q)

24. Compute the value of inductance for a reactance of 15 Ω at 1 GHz. _____(r)

Part V: Change Frequency and Dielectric Constant

25. For the series-matching capacitor:

    a. Calculate the capacitance if the frequency is changed from 1 GHz to 100 MHz: _____(s)

    b. Calculate the length for the matching element if the dielectric constant is 2.1 (Teflon): _____(t)

HINT: See equation 12-21 in the text. The velocity equals "c" divided by the square root of the dielectric constant.

26. For the parallel-matching circuit, start with a 25-Ω load. Change the frequency to 100 MHz and the dielectric constant to 2.1.

    a. Calculate the capacitive reactance and capacitance:

$$\text{(i) } X_C = \underline{\hspace{3cm}}\text{(u)}$$
$$\text{(ii) } C = \underline{\hspace{3cm}}\text{(v)}$$

    b. Calculate the # wavelengths and length for the matching element:

$$\text{(i) Wavelengths } \underline{\hspace{3cm}}\text{(w)}$$
$$\text{(ii) Length } \underline{\hspace{3cm}}\text{(x)}$$

QUESTIONS:

1. Where is the easiest location to read VSWR for the circle you plotted in step 4? Explain.

2. What do all points residing on the VSWR circle you plotted in step 4 represent?

3. In wavelengths, how long is the transmission line section depicted by a full revolution of the VSWR circle on the Smith chart?

4. In your own words, explain how to determine the magnitude and phase of the reflection coefficient at any point along the transmission line.

5. In step 12 and step 23, what was implied when you were asked to add series or shunt values sufficient to make the reflection coefficient equal to zero? Why would this condition be important in a practical context?

6. In Part IV, you matched impedances by adding a capacitor in "shunt." What does "shunt" mean, and why would you use admittance (Y) values to determine the appropriate admittance rather than series impedance values?

7. Refer to Example 12-10 in the text to answer the following question with respect to the steps in Part IV: Under what circumstances would it be advantageous to add a shunt reactive element rather than a series-connected one? Describe all the ways the Smith-chart calculations would have to be modified to perform impedance-matching calculations for shunt-connected reactive elements in contrast to those connected in series.

SMITH CHART MEASUREMENTS USING THE MULTISIM NETWORK ANALYZER

OBJECTIVES:

1. To become acquainted with the use of the Multisim Network Analyzer.
2. To understand impedance measurements.
3. To explore the impedance characteristics of transmission lines and impedance matching for antennas.

REFERENCE:

Refer to Section 12-8 in the text.

EQUIPMENT:

ac voltage source
Multimeter
Virtual resistors, capacitors, inductors
Virtual network analyzer

INTRODUCTION:

The concept of using a Smith chart has been introduced in Chapter 12 in the text. This important impedance calculating tool is now reintroduced using Multisim simulations. Multisim provides a network analyzer instrument that contains the Smith chart analysis as one of its many features. A network analyzer is used to measure the parameters commonly used to characterize circuits or elements that operate at high frequencies. This exercise will focus on the Smith chart and the z-parameter calculations. Z-parameters are the impedance values of a network expressed in its real and imaginary components. Refer to Section 12-8 in the text for additional Smith chart examples and a more detailed examination of its function.

PROCEDURE:

You will be required to make impedance measurements on various resistive, *RC*, and *RL* circuits and transmission lines using the Multisim network analyzer.

Part I: Using the Network Analyzer

1. Begin the exercise by constructing the circuit shown in Figure 27-1. This circuit contains a 50-Ω resistor connected to port 1 (*P1*) of the network analyzer; port 2 (*P2*) is terminated with a 50-Ω resistor. The first circuit being examined by the network analyzer is a simple resistive circuit. This example provides a good starting point for understanding the setup for the network analyzer and how to read the simulation results.

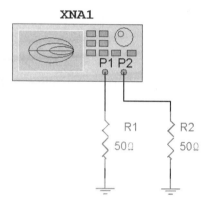

FIGURE 27-1 Multisim network analyzer with test ports resistively terminated.

2. Start the simulation. The impedance calculations performed by the network analyzer are very quick and the start simulation button will quickly reset. Before you look at the test results, predict what you will see. Based on the information you learned in Section 12-8, and the fact that you are testing a resistor, you would expect to see a purely resistive result. Double-click on the network analyzer to open the instrument. You should see a Smith chart similar to the result shown in Figure 27-2.

The Smith chart indicates the following:

$$Z_o = 50\ \Omega$$

Values are normalized to Z_o

$$Z_{11} = 1 + j0$$

The $Z_{11} = 1 + j0$ value indicates that the input impedance for the network being analyzed is purely resistive and its normalized value is 1, which translates to 50 Ω. Recall that the values on a Smith chart are divided by the normalized resistance. Notice the marker on the Smith chart is located at 1.0 on the real axis. The 1.0 translates to 50 Ω, and this value is obtained by multiplying the Smith chart measured resistance of 1 Ω by the characteristic impedance of 50 Ω to obtain the actual resistance measured. In this case the computed resistance is 50 Ω.

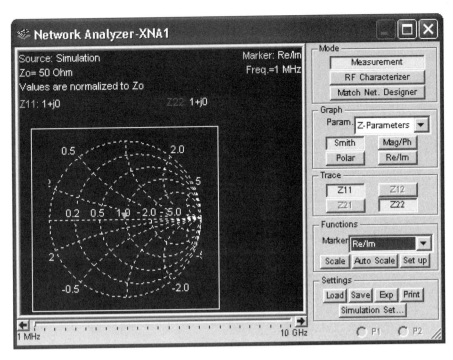

FIGURE 27-2 Smith-chart representation of a resistive load.

The frequency at which this calculation was made is shown in the upper-right corner of the Smith chart screen. In this case a frequency of 1.0 MHz was used. The frequency range for the simulation is shown at the bottom of Figure 27-2 and is adjusted by clicking on the left and right arrows. You can adjust the frequency to see how the impedance values can change through the frequency range. Of course an ideal resistor will not be frequency-dependent. The frequency range, used in the network analysis, is set by clicking on Simulation Set . . . at the bottom of the network analyzer screen. Click the Simulation Set . . . button to check the settings. You will notice the following:

Start Frequency	1 MHz
Stop Frequency	10 GHz
Sweep Type	Decade
Number of points per decade	25
Characteristic impedance Z_o	50 Ω

The start and stop frequency provides control of the frequency range when testing a network. The sweep type can be specified to be plotted in either a decade or linear form, but most of the time use the decade form. The number of points per decade enables the user to control the resolution of the plotted trace displayed, and the characteristic impedance Z_o provides for user control of the normalizing impedance.

3. Change the characteristic impedance of the network analyzer to 75 Ω. This will require changing the characteristic impedance setting for the network analyzer and will also require that the resistor connected to port 2 ($P2$)

be changed to 75 Ω. Change the value of resistor R1 to 75 Ω. Restart the simulation and record the normalized value (Z_{11}); convert the normalized value to actual resistance. Repeat this for the resistor R1 values provided in Table 27-1. Record the measured normalized network impedance and calculate the actual resistance.

TABLE 27-1   Normalized Impedance Values and Actual Resistances for Various Resistance Loads.

R_1 (Ω)	NETWORK IMPEDANCE Z_{11}	RESISTANCE
75		
50		
100		
600		
300		

Part II: Measuring Complex Impedances with the Network Analyzer

The next two Multisim exercises provide examples of using the Multisim network analyzer to compute the impedances of RC and RL networks. These exercises will help you better understand the Smith chart results when analyzing complex impedances.

4. Construct the circuit shown in Figure 27-3. This is a simple RC network of R = 25 Ω and C = 6.4 nF. The network analyzer is set to analyze the frequencies from 1 MHz to 100 MHz. The results of the simulation are shown in Figure 27-4. At 1 MHz, the normalized input impedance to the RC network shows that $Z_{11} = 0.5 - j0.497$. Multiplying these values by the normalized impedance of 50 Ω yields approximately a Z of 25 − j25, which is the expected value for this RC network at 1 MHz. Verify this by calculating the capacitive reactance (X_c) of the 6.4 nF capacitor at 1 MHz. Record your result.

$$X_c = \underline{\hspace{2cm}}$$

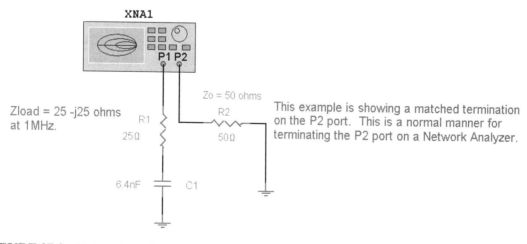

FIGURE 27-3   Network analyzer configuration for series RC network with 25 Ω resistor and 6.4-nF capacitor. Port P2 should always be terminated in characteristic impedance of analyzer.

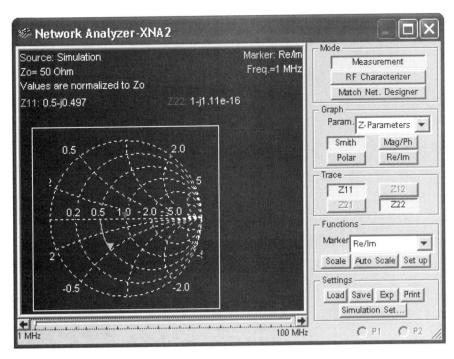

FIGURE 27-4 Smith-chart representation of *RC* network shown in Figure 27-3.

5. Repeat step 4 to measure the impedance of the *RC* networks specified in Table 27-2 for the frequencies listed.

TABLE 27-2   Measured Impedances for Various Combinations of Series Resistance and Capacitance.

R_1 (Ω)	C_1	FREQUENCY	NETWORK IMPEDANCE, Z_{11}
25	6.4 nF	100 kHz	
25	6.4 nF	100 MHz	
10	10 pF	1 MHz	
10	10 pF	100 MHz	
50	50 nF	100 kHz	

6. Next construct the circuit shown in Figure 27-5. This circuit contains a simple *RL* network of $R = 25\ \Omega$ and $L = 4.0\ \mu H$. The network analyzer is set to analyze the frequencies from 1 MHz to 10 GHz. The results of the simulation are shown in Figure 27-6. At 1 MHz, the normalized input impedance to the *RL* network shows that $Z_{11} = 0.5 + j0.5$. Multiplying these values by the normalized impedance of 50 Ω yields $Z = 25 + j25$, which is the expected value for this *RL* network at 1 MHz. Calculate X_L of the 4-μH inductor at 1 MHz. Record your value.

$$X_L = \underline{\qquad\qquad}$$

7. Measure the impedance of the *RL* networks specified in Table 27-3 for the frequencies listed.

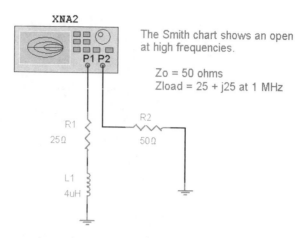

The Smith chart shows an open at high frequencies.

Zo = 50 ohms
Zload = 25 + j25 at 1 MHz

FIGURE 27-5 Network analyzer configuration for series RL network consisting of 25 Ω resistor and 4 μ henry inductor.

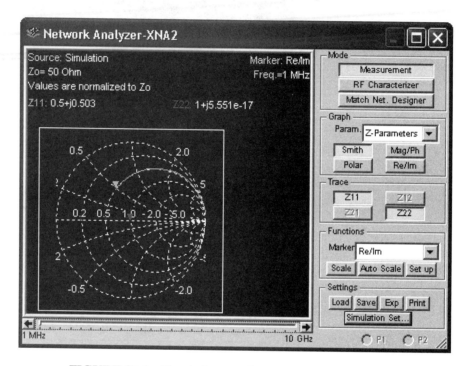

FIGURE 27-6 Simulation results for setup of Figure 27-5.

TABLE 27-3 Measured network Impedances for Series *RL* Circuit at Various Frequencies.

R_1 (Ω)	L_1	FREQUENCY	NETWORK IMPEDANCE, Z_{11}
25	4 μH	10 kHz	
25	4 μH	100 kHz	
25	4 μH	10 MHz	
25	4 μH	100 MHz	
25	4 μH	1 GHz	

8. What observation can you make about the *RL* network of step 7 at low and high frequencies? (*Hint:* Express your observation in terms of the changes in the inductance value.)

Low frequencies:

High frequencies:

QUESTIONS:

1. Explain how a Smith chart displays impedance characteristics. Specifically, what is meant by the "real" and "imaginary" components of impedance, and where would each component appear on the chart?

2. What is the purpose of "normalization" with a Smith chart? How are impedances converted both to and from normalized values to their actual values?

3. Were the results for the simulation performed in step 2 and represented in Figure 27-2 what you expected? Explain.

4. Do you think an ordinary carbon-composition resistor would always show a pure resistance on the Smith chart (or with a network analyzer) at microwave frequencies? Why or why not?

5. Compare the displayed results from Figure 27-4 with those of Figure 27-6. From the Smith chart display alone, are you able to determine the magnitude and type of complex impedance applied to Port P1? What would have to be done in either case to achieve a resistive match?

6. Explain how the RL network behavior changed as a function of frequency (step 8). Why did this occur?

NAME _____

MULTISIM—IMPEDANCE MATCHING

OBJECTIVE:

To use the network analyzer in Multisim to demonstrate impedance matching.

REFERENCE:

Refer to Section 12-10 in the text.

INTRODUCTION:

A network analyzer measures the reflection coefficient and input impedance of an electronic circuit. S parameters are measurements that fully define the transfer characteristics of a microwave circuit. The reflection coefficient for the input port of a circuit is called S_{11}. It has both magnitude and phase information.

PROCEDURE:

Part I: Setting Up the Series Capacitor Circuit

1. Make all of the connections shown in Figure 28-1. The lossless transmission line, "W1," is located in the MISC file. This circuit uses a transmission line and a series capacitor to compensate for the mismatch that is caused by a 25-Ω load positioned at the end of a transmission line having a characteristic impedance of 50 Ω.

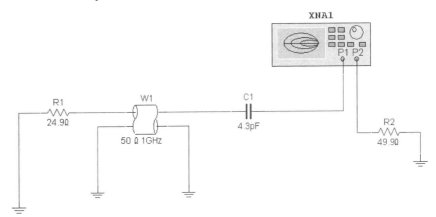

FIGURE 28-1 Initial connections.

2. Click on the transmission line and make the following settings, as shown in Figure 28-2:

> **Nominal Impedance:** 50 Ohm
>
> **Frequency:** 1 GHz
>
> **Normalized Electrical Length:** 0.15

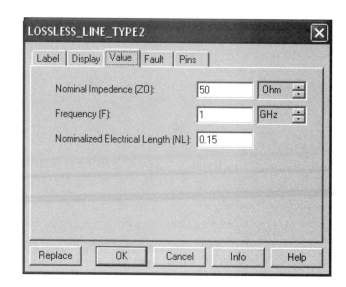

FIGURE 28-2   Transmission line settings.

3. Click on the network analyzer and make the following settings, as shown in Figure 28-3:

> **MODE:** Measurement
>
> **Param:** S-Parameters and Smith
>
> **Marker:** Re/Im

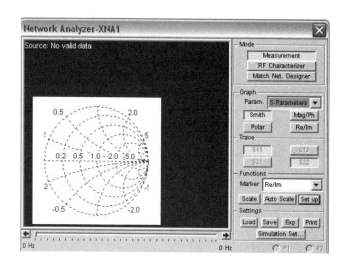

FIGURE 28-3   Network analyzer settings.

4. Set the frequency limits by clicking on "Simulation Set," . . . shown in Figure 28-3. Then make the following settings, as shown on Figure 28-4:

> **Start frequency:** 500 MHz
>
> **Stop frequency:** 1.5 GHz
>
> **Sweep type:** Decade
>
> **Number of points per decade:** 500
>
> **Characteristic Impedance (Zo):** 50 Ohm

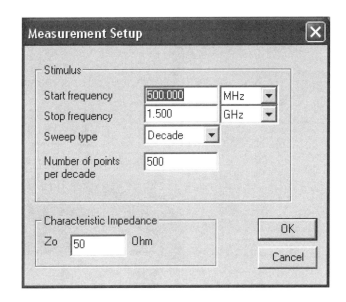

FIGURE 28-4 Frequency range settings.

Part II: Measurements for the Series Capacitor

5. Start the Multisim simulation of the circuit shown in Figure 28-1. Click on the network analyzer. A display similar to that shown in Figure 28-5 should be provided. Move the slider across the screen to change frequency. Note the movement of the marker (red triangle) on the Smith chart. Make the following measurements:

    a. Record the minimum value of S_{11}. _____

    b. Record the frequency when S_{11} is minimum. _____

Ask your instructor to initial your results. _____

6. On the Graph section of the Network Analyzer screen, select Mag/Ph instead of Smith. Under Functions section, select Auto Scale. A display should appear on the Network Analyzer screen that is similar to that shown in Figure 28-6.

7. Move the frequency slide to the left and record the frequency when S_{11} has the following values: 0.05, 0.10, and 0.15. Use Table 28-1.

8. Move the frequency slide to the right of that for minimum S_{11} and record the frequency when S_{11} has the following values: 0.05, 0.10, and 0.15.

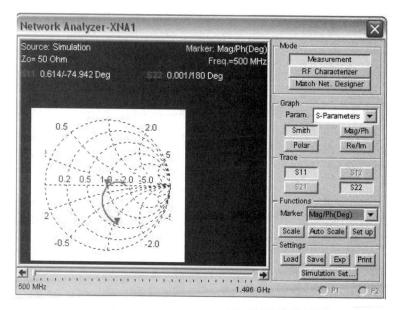

FIGURE 28-5 S_{11} measurements.

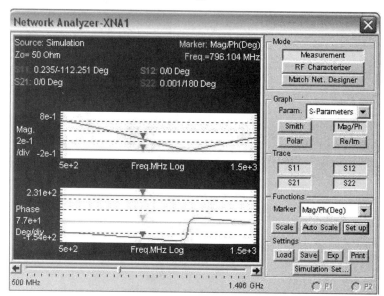

FIGURE 28-6 S_{11} measurements.

TABLE 28-1   Results for Series Capacitor Circuit.

#	S_{11}	FREQUENCY (MHz)	VSWR	BANDWIDTH (MHz)
1	Min:			
2	0.05L			
3	0.10L			
4	0.15L			
5	0.05R			
6	0.10R			
7	0.15R			

9. Use equation 12-27 in your textbook to calculate the VSWR for each of the values of S_{11} recorded in Table 28-1. Note that the magnitude of "S_{11}" is the same as "Γ".

10. Calculate the bandwidth for each value of VSWR.

Part III: Measurements for a Shunt Capacitor

11. Change the circuit as shown in Figure 28-7.

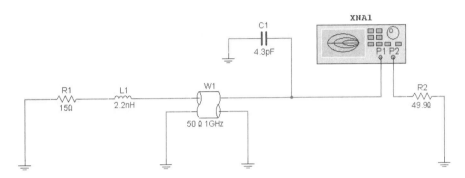

FIGURE 28-7 Connections for a shunt capacitor.

12. For the transmission line remember to change the wavelengths from 0.15 to 0.026.

13. Start the Multisim simulation. A display should appear on the Network Analyzer screen that is similar to that shown in Figure 28-8. Move the slide across the screen to change frequency and record the following:

Minimum S11 _____

Frequency at Min S11 _____

Instructor's initials _____

14. Make the measurements for the shunt capacitor circuit in a manner similar to that used for the series capacitor. Enter the results in Table 28-2.

Part IV: Changing Frequency and Dielectric Constant

15. Make the following changes to the shunt circuit.
    a. For the circuit, shown in Figure 28-7, change R1 and L1 to a single 24.9 Ω resistor.
    b. For the circuit, shown in Figure 28-7, change C1 to 22 pF.
    c. For the transmission line, shown in Figure 28-2, change the frequency from 1 GHz to 100 MHz and the wavelengths from 0.15 to 0.10.
    d. For the Frequency Range, shown in Figure 28-4, change the Start frequency to 50 MHz and the Stop frequency to 150 MHz.
    e. Start the Multisim simulation. A display should appear on the Network Analyzer screen that is similar to that shown in Figure 28-9.

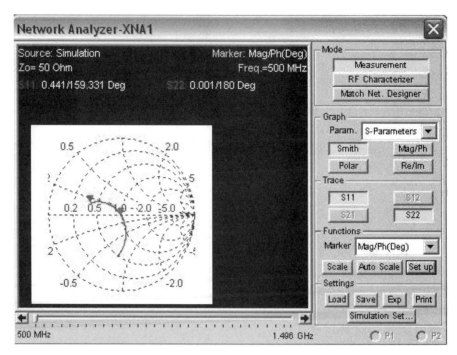

FIGURE 28-8 Measurements for a shunt capacitor.

TABLE 28-2   Results for Shunt Capacitor Circuit.

#	S_{11}	FREQUENCY (MHz)	VSWR	BANDWIDTH (MHz)
1	Min:			
2	0.15L			
3	0.15R			

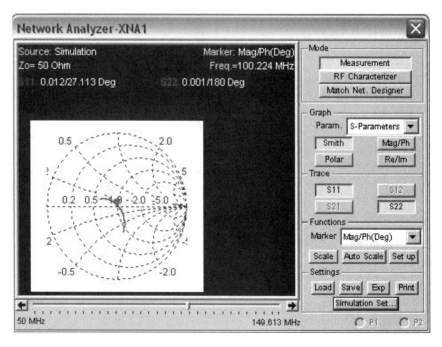

FIGURE 28-9 100 MHz Smith chart for a shunt capacitor.

Part V: Optimizing the Shunt Capacitor Design

16. Replace C1, shown in Figure 28-7, with a virtual capacitor.

17. With the frequency set for minimum S_{11}, record the results in Table 28-3, for the different values of C1. An example of the set-up for the Network Analyzer is shown in Figure 28-10.

TABLE 28-3 Optimizing Results for Shunt Capacitor Circuit.

#	C1	S_{11}	PHASE (DEGREES)	FREQUENCY (MHz)
1	22.0 pF			
2	22.5 pF			
3	23.0 pF			
4	23.5 pF			
5	24.0 pF			
6	23.1 pF			
7	23.2 pF			
8	23.3 pF			

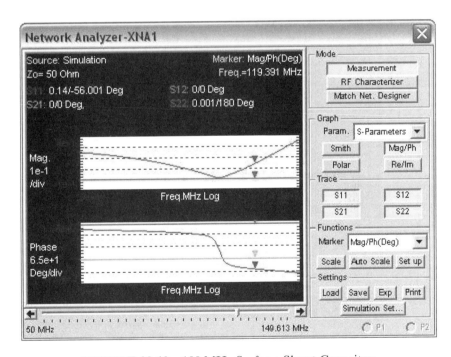

FIGURE 28-10 100 MHz S_{11} for a Shunt Capacitor.

18. Note the large phase change when C1 is changed from 23.0 pF to 23.5 pF. This shows that the capacitance has been changed from a value that is less than ideal to a value greater than ideal.

19. Similarly, note the large phase change when C1 is changed from 23.1 pF to 23.2 pF.

Part VI: Optimizing the Series Capacitor Design

20. Refer to Figure 28-1. Change C1 to a 43.7 pF capacitor.
21. Refer to Figure 28-2. Set the Normalized Electrical Length at 0.15. Set the frequency at 100 MHz.
22. With the frequency set for minimum S_{11}, record the results in Table 28-4, for the different values of C1.

TABLE 28-4   Optimizing Results for Series Capacitor Circuit.

#	C1	S_{11}	PHASE (DEGREES)	FREQUENCY (MHz)
1	43.7 pF			
2	43.8 pF			
3	43.9 pF			
4	44.0 pF			
5	44.1 pF			
6	44.2 pF			
7	44.3 pF			
8	44.4 pF			

23. Note the large phase change when C1 is changed from 50 pF to 51 pF. This shows that the capacitance has been changed from a value that is less than ideal to a value greater than ideal.
24. Similarly, note the large phase change when C1 is changed from 44.0 pF to 44.1 pF.

QUESTIONS:

1. In the simulations shown in Figures 28-1 and 28-7, why is port P2 of the network analyzer connected to a 49 Ω resistor?
2. In Parts V and VI, explain how phase measurements are used to determine optimum capacitor values.
3. In your own words, explain the principal differences between a spectrum analyzer, a scalar network analyzer, and a vector network analyzer.
4. Would you expect an ordinary resistor with leads to appear as a pure resistance on a network analyzer at microwave frequencies? Why or why not?
5. What does the designation "S_{11}" mean in S-Parameter notation? How is it related to familiar quantities such as VSWR and reflection coefficient?
6. What is the main advantage in making network-analyzer measurements of microwave devices and systems with S Parameters as opposed to other parameter sets, such as H, Y, or Z parameters?

SCALAR NETWORK ANALYSIS AND VOLTAGE STANDING-WAVE RATIO (VSWR) MEASUREMENTS

OBJECTIVES:

1. To become acquainted with the basic principles of scalar network analysis through stimulus-response measurements.
2. To examine how component or system parameters such as return loss and insertion loss can be determined through the use of scattering parameters ("S parameters").
3. To examine antenna properties including voltage standing-wave ratio (VSWR), resonance, return loss, and bandwidth.
4. To determine the physical length of a transmission line by making use of the properties of a quarter-wave stub section.

REFERENCE:

Refer to Sections 12-7, 12-10, 14-2, and 15-7 in the text.

EQUIPMENT:

Spectrum analyzer with tracking generator (Rigol DSA-815-TG recommended)

Directional coupler or VSWR bridge (Rigol VB1020 recommended)

Filters: high- and low-pass filters and bandpass filters (recommended: Mini-Circuits NBP-70+ bandpass filter, NHP-1000+ high-pass filter, and NLP-1000+ low-pass filter)

Antenna: Ramsey Electronics LPY-915 recommended

Length of coaxial cable, terminated at one end only

Short lengths of coaxial cable terminated with N male/female connectors

N barrel connector

COMPONENTS:

None

INTRODUCTION:

This experiment is intended to acquaint you with some principles of network analysis and with the quantities associated with matched and mismatched impedances. A *network analyzer* is a piece of test equipment used to characterize the impedance characteristics of devices under test (DUTs). DUTs can be circuits such as amplifiers or filters, loads such as antennas, or individual components such as transistors or diodes. Network analyzers allow users to characterize circuit behavior completely by measuring impedance-related characteristics at the input and output ports of the DUTs only. Essentially, the DUTs are treated as "black boxes" because, once the relevant parameters have been measured and calculations made, device or circuit behavior in any external environment can be predicted without regard to the internal configuration of the DUT.

Network analyzers measure the transmission and reflection characteristics of signals applied to the inputs and emerging from the outputs of DUTs. Recall from electrical fundamentals that impedance can consist of both resistive (real) and reactive (imaginary) elements. Impedances with both elements present are said to be *complex*. Complex impedances can be expressed either in rectangular form ($Z = R \pm jX$), suitable for analysis with the Smith chart, or in polar form ($Z \angle \theta$), where Z represents the magnitude and θ the phase angle created by the presence of reactances. (In a series circuit with reactance, θ represents the number of electrical degrees of phase shift between source voltage and circuit current as seen by the source.) A vector network analyzer provides a full picture of device behavior by capturing impedance magnitude and phase characteristics at all DUT inputs and outputs. A scalar network analyzer measures only the magnitude of impedance, not its phase angle. A number of important performance parameters can be determined from magnitude alone, among them insertion loss, return loss, and reflection coefficient.

A spectrum analyzer can function as a scalar network analyzer if it is equipped with a directional coupler and a tracking generator. As described in Section 15-7 in the text, a directional coupler isolates transmitted (forward) from reflected (reverse) components of energy within a transmission line and allows for measurement of each component separately. Scalar network analyzers will have directional couplers integrated within them, as will some spectrum analyzers with "stimulus-response" capability. Alternatively, the coupler (sometimes referred to as a VSWR bridge) may be connected externally between the tracking generator output and the analyzer RF input. The tracking generator is a radio-frequency source whose output frequency is tied to the frequency display of the analyzer. As the analyzer display sweeps from left to right, the generator output increases from the lower start frequency represented by the left side of the display to the higher stop frequency on the right side.

Many impedance-related quantities of interest can be described in terms of S parameters. As covered in more detail in Section 12-10 in the text, S parameters describe all voltages incident to and reflected from all DUT ports in the form of a "scattering matrix" ("S" for "scattering"), thus modeling fully the behavior of signals applied to the DUT. The term comes about because signals incident to any network port can be thought of as scattering in all possible directions as they propagate through the network. The signal-scattering effect within a DUT is conceptually similar to what happens when a cue ball hits a full rack of billiard balls and causes them to travel in widely scattered directions. Similarly, signals incident to any port can also be reflected back through that port because of impedance mismatches, or signals may be output from or reflected by any other port. These measurements produce results that can effectively be represented as a system of multiple equations with multiple unknowns and whose variables represent voltages that can be determined when the equation coefficients are arranged in matrix form and solved. In S parameter notation, both incident and reflected signals are shown as subscripts, where the first subscript is the exit port and the second subscript the incident port. Therefore, a network consisting of ports numbered 1 and 2 will have reflection coefficients denoted by S_{11} or S_{22} (that is, a signal incident to port 1 or 2, respectively, is also reflected back to the port from which it was incident); similarly, transmission coefficients are represented as S_{21} or S_{12}, depending on the incident and exit ports involved.

A significant advantage of S parameters, and a contrast with the way measurements must be made with other possible parameter sets, is that S parameter measurements are always determined with ports resistively terminated rather than with ports open- or short-circuited. True opens or shorts can be difficult to achieve, particularly at radio frequencies. For example, a circuit that appears open at low frequencies can act as a reactive circuit at high frequencies. Such conditions could cause the DUT to oscillate and possibly self-destruct. S parameters, on the other hand, are determined with all ports resistively terminated. In a practical setup, the DUT is often placed in a transmission line between a 50-Ω source and a resistive load, thus making accurate measurements possible because oscillations are unlikely to occur. Also, because incident and reflected traveling waves do not vary in magnitude at points along a lossless transmission line, S parameters can be measured on DUTs located some distance from the measurement transducers (i.e., the source and reflection measurement points) as long as low-loss transmission lines are used and as long as residual transmission line effects can somehow be "normalized out" or eliminated from subsequent calculations.

PROCEDURE:

Part I. Two-Port Measurements

1. Follow the steps shown below to perform a two-port insertion loss measurement of the bandpass filter. (The term *two-port* refers to the fact that both the input and output port of the filter are connected at the same time to the tracking-generator output and to the RF input ports of the analyzer.) Two-port insertion-loss measurements accurately quantify the amount of loss or gain a signal will incur as it passes through a device. In S-parameter terms, insertion loss is referred to as an S_{21} measurement. (These procedure steps assume use of the Rigol DSA-815-TG analyzer and VB1020 VSWR bridge. The procedure will be similar for other analyzer/VSWR bridge combinations.)

    a. Set the analyzer up to make a two port insertion loss measurement by first pressing the Freq function key.

    b. If using the Mini-Circuits NBP-70+ bandpass filter, set the start and stop frequencies as follows. (Your instructor will advise as to the start and stop frequencies for other types of bandpass filters):

        a. Press **Start Freq,** 50, **MHz**

        b. Press **Stop Freq,** 100, **MHz**

    c. Press the TG function key to access the tracking generator menu. Set TG Level to 0 dBm.

    d. Normalize the tracking generator output by connecting the tracking generator output directly to the RF input. The purpose of normalization is twofold: First, it compensates for any amplitude variations in tracking generator output over the frequency range of interest.

Second, normalization allows for compensation for the loss associated with the devices (i.e., adapters, cables) that connect the analyzer to the device or assembly being tested. Otherwise, the loss introduced by these connecting devices is added to the loss of the device under test. For this reason, you should use the same coaxial cables you intend to use for the next steps to perform the normalization. An adaptor may be needed to connect two lengths of cable together temporarily. Normalization must be performed over the intended frequency range before any two-port insertion loss measurement is made. To perform the normalization, turn on the tracking generator by pressing the TG soft key and press the Normalize soft key.

e. After the normalization is completed, determine the filter characteristics by connecting the filter between the tracking generator output and RF input connections of the analyzer. The resulting trace displays the insertion loss of the filter over the frequency range entered in step 1(c). It should be similar to the display of Figure 29-1. Press **Marker** and turn the knob to move the cursor along the displayed trace.

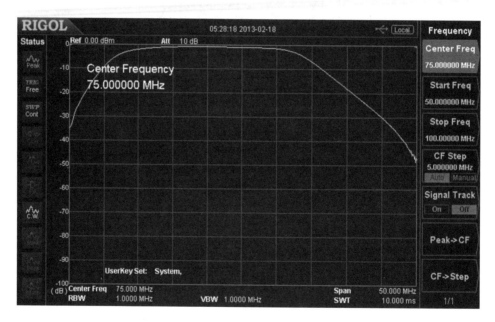

FIGURE 29-1 Response curve of bandpass filter.

Insertion loss is a measure of attenuation caused by a device placed within in a signal path. For a bandpass filter, the insertion loss should be minimum in the passband (the frequency region of minimum attenuation) and maximum in the stopband (the frequency region of maximum attenuation). What is the minimum insertion loss of this filter in the passband?

_____ dB

What is the minimum insertion loss of the filter in the stopband?

_____ dB

The *critical frequency* of a filter is generally defined as the frequency at which the filter output amplitude decreases by 3 dB from the maximum output seen in the passband. Because this is a bandpass filter there will be two critical frequencies, f_1 and f_2, in which f_1 is below and f_2 is above the center frequency. You can determine the critical frequencies by moving the marker and observing the points at which the displayed signal level is reduced by 3 dB from the minimum attenuation (hence maximum signal level) in the passband. The critical frequencies are those on either side of the passband where the signal has been reduced by 3 dB from the level in the passband.

What are the critical frequencies, f_1 and f_2?

f_1: _____ MHz; f_2: _____ MHz

What is the filter bandwidth $(f_2 - f_1)$: _____ MHz

f. Confirm your calculated bandwidth and critical frequencies with those the analyzer determines automatically. Press the Marker Fctn key and the N dB BW soft key. Note also that, by changing the decibel value specified for N dB BW, the bandwidths at attenuations other than those represented by the –3-dB points can also be determined.

g. Refer to the manufacturer's specification data to compare your measurement with the published specifications for return loss and 3-dB bandwidth. Virtually all manufacturers now have web-based tools for looking up specification data. For example, data for all Mini-Circuits modules can be found at http://www.mini-circuits.com.

Part II. One-Port Measurements

2. We will now perform a return loss measurement on the filters. *Return loss* is a measure of the reflection characteristics of a DUT and is derived from the ratio of the reflected voltage to the incident voltage at any port of the DUT. This ratio, known as the reflection coefficient, is represented by the Greek letter gamma (Γ) and is a complex number, meaning it has both magnitude and phase information. The magnitude of the reflection coefficient is represented by the Greek letter rho (ρ), and the return loss, RL, is related to ρ as $RL = -20 \log \rho$. In S-parameter terms, return loss and reflection coefficient are both S_{11} measurements.

The return loss measurement described in this step is an example of a "one-port" test because only one of two ports of the DUT is connected to the analyzer. The VSWR bridge, pictured in Figure 29-2, is a directional coupler that applies a signal from the tracking generator (a "stimulus") to the DUT as it isolates the reflected signal energy caused by an impedance mismatch and applies that reflected energy to the analyzer input (the "response").

Use either the high-pass or low-pass filters to perform the following return loss tests.

a. Set up the analyzer to make a one-port return loss measurement by installing the VSWR bridge onto the front of the instrument, as shown in Figure 29-3.

b. Set the start and stop frequencies so that both the passband and stopband behaviors of the filters are represented. As an example, if either

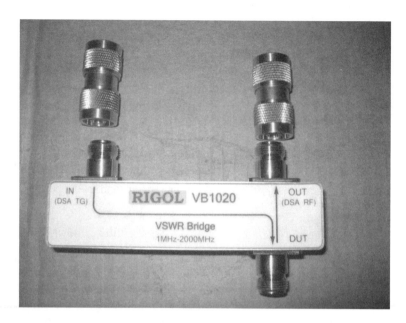

FIGURE 29-2 Rigol VSWR bridge.

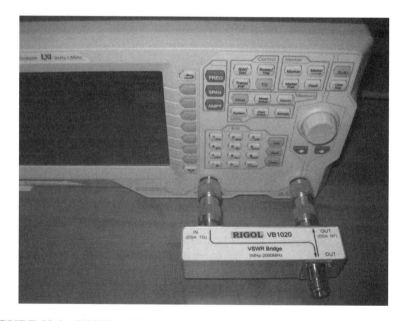

FIGURE 29-3 VSWR bridge installed onto front of Rigol spectrum analyzer.

the Mini-Circuits NHP-1000+ high-pass filter or NLP-1000+ low-pass filters is used, the start frequency can be set for 850 MHz, and the stop frequency to 1.1 GHz.

c. Confirm that the tracking generator level is set for 0 dBm by pressing the TG function key; adjust as necessary. Turn on the tracking generator and press the Normalize soft key. Turn normalization on. You should see the line on the screen jump up to the reference level because the normalization routine has effectively subtracted out the attenuation presented by the VSWR bridge.

d. Perform the test: connect the filter to the DUT port of the VSWR bridge and connect the 50-Ω termination to the unused filter port. Observe the display. It will be similar to the one in Figure 29-4 for the NHP-1000+ high-pass filter. Use the markers to measure the return loss at any point by pressing **Marker**. Use the knob to place the marker at a frequency of interest.

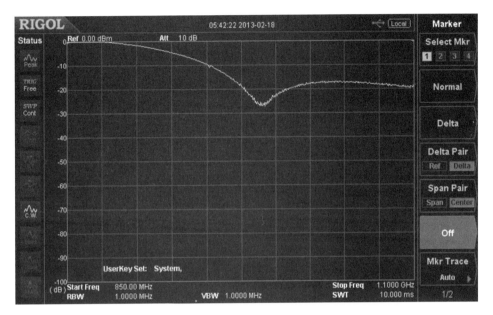

FIGURE 29-4   Response curve for NHP-1000+ high-pass filter.

Figure 29-4 is a display of return loss in decibels as a function of frequency. Recall from Section 12-7 in the text that return loss is defined as the ratio of forward power to reflected power. Like VSWR and reflection coefficient, return loss is an expression of the degree of impedance mismatch between source and load. Because a good impedance match would produce little or no reflected energy, a high return loss would be equivalent to a low VSWR or a reflection coefficient close to or equal to zero and would, therefore, be associated with matched impedances. For a filter, the match would be best, therefore the return loss highest, in the passband because most of the incident energy is passed through the filter to the load. Conversely, a low return loss indicates that a great deal of energy is returned to the source, which is consistent with filter behavior in the stopband.

What is the return loss in the passband? _____ dB.   The SWR: ____

What is the return loss in the stopband? _____ dB.   The SWR: ____

What is the critical frequency? _____ MHz

3. Another very useful application of a return loss measurement is to detect problems in an antenna feedline system or in the antenna itself. A portion of the incident power will be reflected back to the source from each transmission line fault as well as from the antenna. Using the antenna supplied and coaxial cable, perform a return loss measurement following the procedure shown in step 2 above. (Note that you will have to renormalize the analyzer for the frequency range for which the antenna is designed, which should include its resonant frequency.) After the normalization is complete, connect the antenna to the DUT port of the VSWR bridge. On the grid below, make a sketch of the trace and indicate the frequency at which the antenna is most nearly resonant.

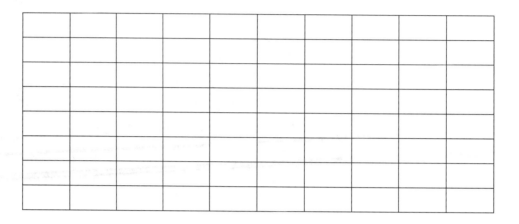

4. Antenna VSWR can be determined from return loss. First, the magnitude of the reflection coefficient, previously defined as ρ, can be calculated from the return loss in decibels as follows:

$$\rho = 10^{\frac{-RL(dB)}{20}},$$

where $-RL(dB)$ is the return-loss value from the spectrum analyzer display. VSWR can then be calculated from ρ with the following:

$$VSWR = \frac{1 + \rho}{1 - \rho}.$$

Calculate the VSWR associated with your antenna at the frequency at which it is most resonant.

5. The behavior of a quarter-wavelength section of open-ended coaxial cable can be used to determine its physical length. Recall from Section 12-6 in the text that an open circuit appears as a short circuit to a source located one-quarter wavelength away because voltage and current on the line undergo a 180° phase inversion: a high-voltage/low-current relationship at the open appears as a low voltage and high current relationship at the generator (see Figure 29-5). The line thus acts as a notch filter to the quarter-wavelength frequency, and this behavior, visible on the spectrum analyzer, coupled with knowledge of the velocity factor of the line, allows determination of physical length.

   a. Remove the VSWR bridge if it is still connected. Connect the tracking generator output to the RF input of the analyzer with two short cables and a tee connector, as shown in Figure 29-6.

   b. Connect the open-ended cable of unknown length to the tee.

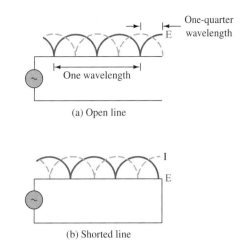

(a) Open line

(b) Shorted line

FIGURE 29-5   Standing waves of voltage and current.

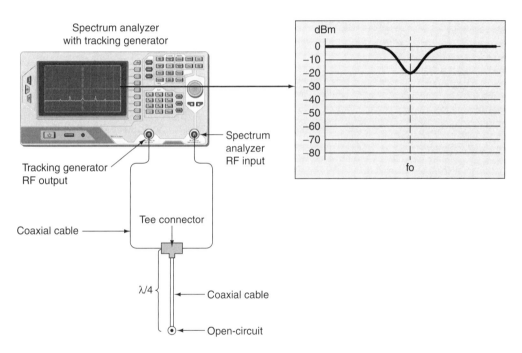

FIGURE 29-6   Test setup for determining physical length of coaxial cable.

c. Press the Freq function keys and select the start and stop frequencies to be within the range where the cable is expected to be one-quarter wavelength. (For a 20-ft long cable, this frequency would be in the area of 10 MHz, for example; for a 1-ft long cable, it would be around 150 MHz.)

d. Adjust the stop frequency until the first dip is clearly visible, as shown in Figure 29-7. Use the marker to determine the exact frequency of the dip.

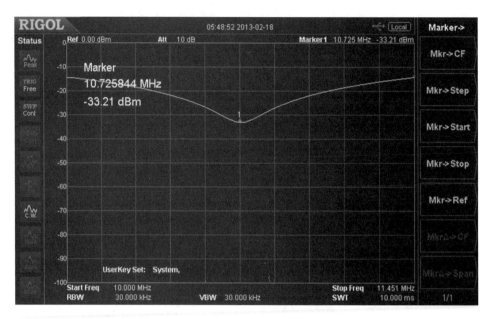

FIGURE 29-7 Frequency sweep showing first notch corresponding to quarter-wavelength phase inversion of voltage and current on coaxial cable.

e. Determine or estimate the velocity factor of the unknown-length cable. This information can be found from manufacturer data sheets. For RG-8U type cable, a typical value for velocity factor would be 0.66.

f. The physical length of the cable is equal to its electrical quarter wavelength, $\lambda/4$, multiplied by the velocity factor (expressed as a decimal quantity, not as a percentage). For lengths in feet, the quarter wavelength is calculated as follows:

$$\lambda/4 = \frac{245.9}{f_{MHz}} \times V_f,$$

where f_{MHz} is the frequency of the first dip determined in step 5(d) and V_f is the velocity factor determined in step 5(e).

QUESTIONS:

1. Compare your displayed frequency responses from step 2(d) with the return-loss graphs provided in the manufacturer data sheets. How closely do your measured return loss values agree with the manufacturer specifications? How does the manufacturer's representation of return loss differ from the display from the analyzer?

2. Explain the relationship between return loss, reflection coefficient, and VSWR. Specifically, which of each of these values is indicative of a matched source and load? And which is indicative of an open- or short-circuited condition?

3. Describe a practical application for the cable-length determination technique outlined in step 5.

4. Describe how you would modify the procedure of step 5 to determine experimentally the velocity factor of a transmission line of known length.

5. In steps 5(c) and 5(d) you increased the frequency supplied by the tracking generator until you saw the first instance at which the signal appeared attenuated (i.e., the first "dip"). Would you expect to see additional dips if you continued to increase the frequency? At what frequencies would these additional dips occur?

6. Can an antenna function effectively at frequencies other than the frequency at which it is resonant? Explain. Also explain how this notion ties into the notion of antenna bandwidth and VSWR.

7. Can an antenna have more than one resonant frequency? Explain.

8. You are in a job interview. Explain concisely what types of instruments you would use and what measurements you would make to verify the overall health of a newly installed antenna system and its associated feedline.

NAME _____

ANTENNA POLAR PLOTS AND GAIN CALCULATIONS

OBJECTIVES:

1. To determine the relationship between received signal strength and distance between two antennas.
2. To plot the radiation pattern of antennas with and without reflectors.
3. To calculate the gain and beamwidth of an antenna with and without a corner reflector.

REFERENCE:

Refer to Sections 14-2 and 14-6 in the text.

EQUIPMENT:

Spectrum analyzer with tracking generator capable of operation at 915 MHz or higher (Rigol DSA-815-TG recommended)

2 Yagi antennas (Ramsey Electronics RPY915 Yagi antennas recommended)

2 antenna mounts consisting of wooden bases and 4-ft long dowels

2 antenna reflectors

2 lengths of high-quality, 50-Ω coaxial cable, each between 6 and 10 ft, with appropriate connectors or adaptors

4 each 1-m-long measuring sticks or equivalent

COMPONENTS:

Miscellaneous N and BNC male/female and between-series adaptors as needed

Reflector mounting hardware

INTRODUCTION:

The purpose of this experiment is to demonstrate the radiation properties of antennas. An antenna is a transducer that converts electrical energy to electromagnetic energy at the transmitter and that performs the converse transformation at the receiver. The simplest type of antenna to construct is the half-wave dipole, which essentially consists of two quarter-wavelength conductors oriented 180° from each other. This orientation allows energy that would otherwise be confined within a section of two-conductor transmission line instead to be launched into free space and to propagate away from the antenna at the speed of light. In general, antennas are complex-impedance loads; that is, they appear to a source as both resistive and reactive elements connected in series. However, at the frequency at which the antenna is electrically one-half wavelength, it is said to be *resonant* because it presents a purely resistive load to the source. If the antenna impedance at resonance is equal to the source impedance and to the characteristic impedance of its associated transmission line, then the antenna will radiate all the energy supplied by the source into the surrounding environment. Any impedance mismatch will cause a portion of the incident energy to be reflected back to the source.

Though inherently passive devices, practical antennas exhibit gain by focusing energy in some directions at the expense of others. The effect is similar to that produced by a reflector placed behind a flashlight bulb. The light beam appears brighter than it would have otherwise because the energy that would have traveled behind the bulb without the reflector present instead adds to the light produced directly and travels in the forward direction. Gain is effectively achieved even though additional electrical energy has not been applied to make the bulb brighter. Antennas achieve gain in a similar fashion, either with metallic reflectors placed behind the driven element or with systems of conductors arranged such that currents induced within each conductor produce electromagnetic energy that is in phase with that radiated by adjacent conductors and that is constructively additive in the desired direction. Some antennas employ both techniques to achieve very high gains.

Gain antennas are inherently directional because of the focusing provided by the reflector or array of conductors. A measure of the directive effect is expressed in degree terms as *beamwidth,* which is generally defined as the points on either side of the direction of maximum radiation at which the signal has been reduced by 3 dB. Along with gain and other properties, beamwidth is best visualized through the use of polar radiation plots, which are cross-sectional views of the three-dimensional radiation properties of antennas. A full picture requires both horizontal (azimuth) and vertical (elevation) polar plots. In this experiment we will produce polar plots in the azimuth plane only for a gain antenna both with and without a reflector in place.

There are several techniques for determining antenna gain. The most accurate determinations require reference antennas calibrated in purpose-built facilities designed to minimize reflections caused by nearby objects. Indoor anechoic chambers are often used for close-range characterizations, particularly of antennas operating at microwave frequencies. Other, larger antennas may require the use of outdoor antenna ranges, which are literally wide-open spaces devoid of metallic objects and consisting of wooden support structures to allow devices under test to be positioned high enough to minimize the effects of ground reflections.

The method to be used in this experiment employs two antennas of the same type to determine gain based on received signal levels at two locations. The transmitted and received powers, P_T and P_R, can be measured and the gain calculated with the following formula, which is a modification of the free-space path-loss relationship to solve for gain:

$$G = \frac{4\pi r}{\lambda} \sqrt{\frac{P_R}{P_T}},$$

where λ is the signal wavelength and r is the distance between the two antennas. (Both quantities should be expressed in the same units, which will be centimeters for the frequency used in this experiment.) The result gives the gain, G, as a power ratio referenced to an isotropic point source. To convert the gain ratio to decibel form, use the expression

$$G_{(dB)} = 10 \log G.$$

Though applicable in principle at all frequencies, the identical antennas method of gain determination embodied in the following procedure steps produces the best experimental results with horn antennas at microwave frequencies. The short wavelengths and small antenna dimensions encountered there would cause the signal source to act most like a theoretical *point-source radiator,* in which the energy power density would decrease in an inverse-square relationship with distance as it spreads out spherically and propagates equally in all directions away from the radiator. The inverse-square relationship and path loss formula both hold true in the far field, but some variation in experimental results and deviation from theory may be encountered in this experiment because of reflections from nearby objects and because the frequency specified, 915 MHz, requires the use of antennas whose dimensions are significantly larger than those for microwave applications. Dipoles and other types of antennas encountered at lower frequencies, including the Yagi antenna used in this experiment, require substantially more involved models of their radiation characteristics than that predicted by the free-space model. However, even with the variations from the ideal situation presented by equipment and frequency limitations, we can still use the free-space path loss formula to confirm experimentally both the 6-dB drop in signal level expected when the distance between antennas is doubled as well as to see the focusing effect provided by a metallic reflector.

PROCEDURE:

Your instructor will provide materials sufficient to create the setup pictured in Figure 30-1.

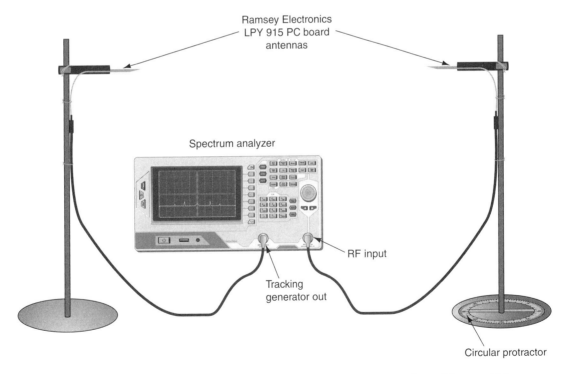

FIGURE 30-1 Experimental setup for determining antenna gain and beamwidth.

The Rigol DSA-815 spectrum analyzer with the "TG" option has a tracking generator that will serve as the 915-MHz frequency source for this experiment. The analyzer will display the received power in dBm at specified antenna separations.

1. Determine the start of the far field (radiation field) for the antenna by calculating the wavelength, λ, at 915 MHz, determining the longest dimension of the antenna, and using the appropriate form of Equation 14-1 from the text to determine the far-field region, R_{ff}:

   a. $\lambda = \dfrac{c}{f} = \dfrac{3 \times 10^8 \, \text{m/s}}{915 \times 10^6 \, \text{Hz}} = \underline{\hspace{1cm}}$ cm

   b. $D = \underline{\hspace{1cm}}$ cm

   c. $R_{ff} = \underline{\hspace{1cm}}$ cm

2. Set up the equipment as shown in Figure 30-1. Initially, the antennas should be spaced 50 cm apart and mounted such that the printed-circuit etched conductors are parallel with the work surface and approximately 4 ft above it. Initially, no reflectors will be used. The best results will be obtained with an open space, so make sure no objects or people are in the immediate vicinity of your setup. You will need to position the analyzer such that it can be connected to the antennas and the display seen without anyone being in the immediate area. If practical, repeat the experiment outdoors (such as in an empty parking lot or athletic field) to verify your results.

3. Using one of the coaxial cables, attach the left-hand antenna to the tracking generator output of the analyzer. Use the other coaxial cable to attach the right-hand antenna to the analyzer RF Input. Power on the analyzer and set its center frequency to 915 MHz. Set the span to zero span by pressing **SPAN** and selecting the zero span soft key. This selection causes the tracking generator to produce a constant-frequency output. Next select the tracking generator menu by pressing TG. On the tracking generator menu set the generator output to 0 dBm, and turn on the generator by pressing the TG on/off soft key. The analyzer will display a horizontal line depicting the received signal level. Press the MARKER key to display the received level in dBm.

4. Observe the received signal level as you get closer to and then farther away from the experimental setup. Place your hand near the antennas and in the space between them to see the effect of obstructions on received signal level.

5. Again make sure that the surrounding area is free of obstructions and close-by surfaces or objects that can create reflections. Aim the antennas directly at each other and adjust the receive antenna for highest signal level. Record the received signal strength in dBm at each distance shown in Table 30-1.

TABLE 30-1   Received Signal Level as a Function of Distance at 915 MHz for Antennas without Reflectors.

DISTANCE (METERS)	RECEIVED SIGNAL LEVEL (dBm)
0.5	
1	
1.5	
2	
3	
4	

You should see a constant reduction as the antennas are separated from each other. Compare the levels at 1 m and 2 m. By how many decibels is the signal at 2 m reduced from the level at 1 m? _____ At 3 m from 2 m? _____.

6. Separate the antennas from each other by 1 m and ensure that the antennas are aligned correctly. Orient the base of the receive antenna such that its azimuth position indicator shows zero degrees when the antennas are directly facing each other. In Table 30-2, record the received signal level (RSL) at zero degrees in the first row. This RSL is your reference level. Then, rotate the receive antenna in 10° increments and record the RSL at each point in the appropriate row. Calculate the power ratio by subtracting the RSL you recorded for each position from the reference RSL at 0°.

TABLE 30-2   Recorded Receive Signal Levels and Calculated Power Ratios in the Azimuth Plane for Antenna without Reflector.

ANTENNA AZIMUTH (DEGREES)	RECEIVED SIGNAL LEVEL (dBm)	POWER RATIO (dB)	ANTENNA AZIMUTH (DEGREES)	RECEIVED SIGNAL LEVEL (dBm)	POWER RATIO (dB)
0		0	180		
10			190		
20			200		
30			210		
40			220		
50			230		
60			240		
70			250		
80			260		
90			270		
100			280		
110			290		
120			300		
130			310		
140			320		
150			330		
160			340		
170			350		

7. Using the power ratios you calculated and recorded in Table 30-2, plot the antenna radiation pattern using the polar plot diagram shown in Figure 30-2.

8. We will now see the enhancement of gain produced when a reflector is placed behind each antenna. Your instructor will provide two reflectors and hardware necessary to secure the reflectors onto the masts. After the reflectors are in place, again separate the antennas, initially by 0.5 m and then to the distances shown in Table 30-3, and record the signal level at each point.

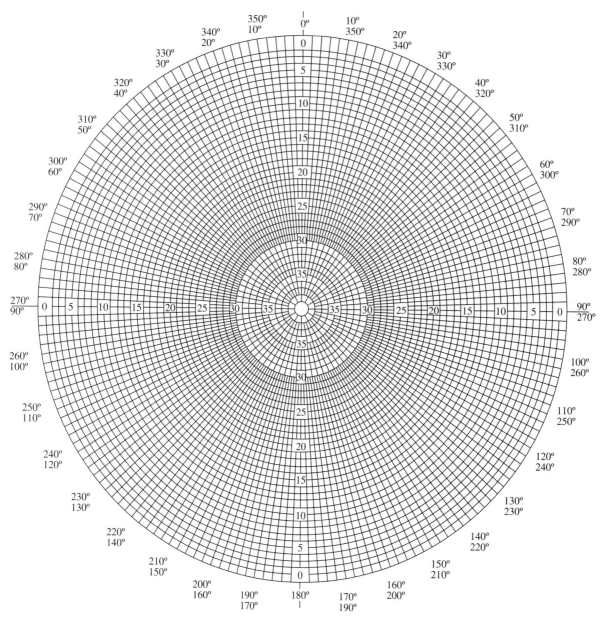

FIGURE 30-2 Radiation pattern (azimuth plane) for antenna without reflectors.

TABLE 30-3   Received Signal Level as a Function of Distance at 915 MHz for Antennas with Reflectors.

DISTANCE (METERS)	RECEIVED SIGNAL LEVEL (dBm)
0.5	
1	
1.5	
2	
3	
4	

9. Compare the results from the preceding table with the results recorded from step 5. Are the received signal levels higher? Again compare the levels at 1 m and 2 m and at 2 m and 3 m. How closely did the reduction in signal level compare with the expected reduction of 6 dB when distance was doubled?

10. Repeat the procedure outlined in step 6 to determine the radiation pattern of the antennas with reflectors. Record your measurements and calculations in Table 30-4.

TABLE 30-4   Recorded Receive Signal Levels and Calculated Power Ratios in the Azimuth Plane for Antennas with Reflectors.

ANTENNA AZIMUTH (DEGREES)	RECEIVED SIGNAL LEVEL (dBm)	POWER RATIO (dB)	ANTENNA AZIMUTH (DEGREES)	RECEIVED SIGNAL LEVEL (dBm)	POWER RATIO (dB)
0		0	180		
10			190		
20			200		
30			210		
40			220		
50			230		
60			240		
70			250		
80			260		
90			270		
100			280		
110			290		
120			300		
130			310		
140			320		
150			330		
160			340		
170			350		

11. Using your calculated power ratios from Table 30-4, plot the radiation pattern of the antennas with reflectors using the polar plot diagram shown in Figure 30-3.

12. As described the introduction, the antenna beamwidth is defined as the number of degrees between the points, one on each side of the point of maximum radiation, where the signal level has fallen by 3 dB from the maximum (represented by the signal strength at 0° azimuth). Determine the beamwidth for the antennas both without and with the reflectors.

Beamwidth without reflector: _____

Beamwidth with reflector: _____

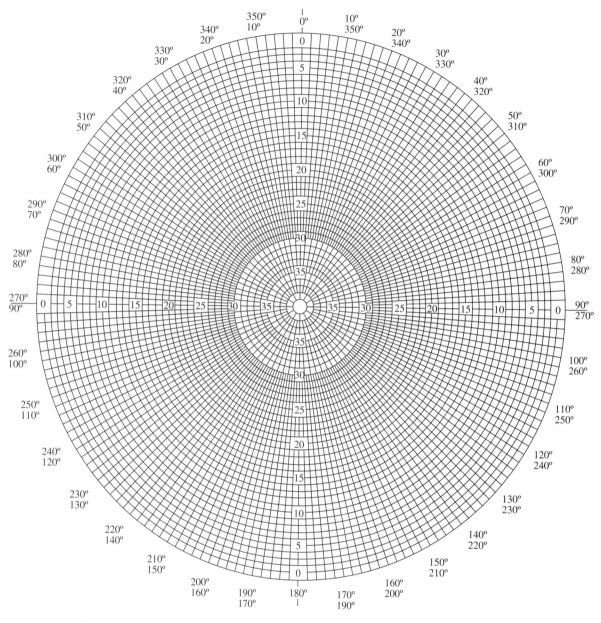

FIGURE 30-3 Radiation pattern (azimuth plane) for antenna with reflectors.

13. We will first use the data you collected in step 8 and recorded in Table 30-4 to determine antenna gain. First, we will need to determine the transmit power, P_T. Temporarily disconnect the transmit antenna from the long coaxial cable, and connect the cable from the tracking generator output to the RF input of the analyzer. (Be sure this is the same cable you used for the earlier procedure steps.) Ensure that the transmit frequency is still 915 MHz, that the span is set for zero span, and that the tracking generator is turned on and the output set for 0 dBm. The displayed power level will be less than 0 dBm because of cable losses. This power, when delivered to the antenna, is the transmit power, P_T, that will be used in the

gain formula given in the introduction. Record the transmit power here: $P_T = -\rule{1cm}{0.4pt}$ dBm.

14. The signal level recorded at 1 m in step 8 will be the receive power, P_R. Because both P_T and P_R are expressed in decibel units, we will first need to convert these values to a power ratio for use in the gain calculation. Subtract P_R in decibels from the value for P_T recorded in step 13. Remember that you are dealing with reductions in signal level (negative gains), so all the negative signs are important. As an example, if your P_T is -8 dBm, and your P_R is -24 dBm, then the calculation would appear as follows:

$$\text{Power Ratio (dB)} = P_R\,(\text{dB}) - P_T\,(\text{dB})$$
$$= -24 - (-8) = -16\,\text{dB}$$

The numbers used in this example are typical, but you should substitute the values you determined experimentally. Notice the negative number in the result—it must be included in the calculations performed in the next step. Record your power ratio in dB here: $\rule{1cm}{0.4pt}$ dB

15. Use the following expression to calculate the power ratio:

$$\frac{P_R}{P_T} = 10^{(\text{POWER RATIO (dB)}/10)}$$

For the example results from the previous step, the preceding calculation would appear as follows:

$$\frac{P_R}{P_T} = 10^{(-16/10)} = 10^{-1.6} = 0.025$$

Use the power ratio in decibels you calculated in step 14 to determine the power ratio and record your result here:

$$\frac{P_R}{P_T} = \rule{1.5cm}{0.4pt}$$

16. Use your calculated wavelength from step 1, the distance between antennas (1 m) for the value of r in the gain calculation formula, and the power ratio determined in the previous step to calculate the antenna gain. Remember that r and λ must be in the same units for this calculation:

$$G = \frac{4\pi r}{\lambda}\sqrt{\frac{P_R}{P_T}} = \rule{2.5cm}{0.4pt}$$

17. Convert the gain figure to decibel gain at one meter as follows:

$$G\,(\text{dB}) = 10\log G = \rule{1.5cm}{0.4pt}\,\text{dB}.$$

18. Repeat steps 14 through 17 substituting the received power level you recorded at 2 m in step 8. How closely do the computed gains at 1 and 2 m agree with each other?

19. Repeat the gain calculations for the measurements you recorded at 1 m and 2 m in step 5. What was the gain of the antennas without reflectors behind them?

dB

QUESTIONS:

1. How closely did the results you obtained in step 5 compare with the 6-dB reduction in power predicted by the inverse-square law? Explain why 6 dB is the expected power reduction.

2. What would be the expected reduction in amplitude (induced voltage) in the receive antenna when the distance between the antennas is doubled?

3. What factors in your experimental results or setup would explain any deviations from theory in steps 5 and 9?

4. Would you expect to see a constant reduction in signal strength between 25 cm and 50 cm? Between 50 cm and 1 m? Why or why not? Would the gain calculations be affected if the received power levels at 25 cm were used rather than those made at 1 m?

5. Discuss how the radiation plot shown in Figure 30-3 for the antennas with reflectors shows increased gain over that depicted in Figure 30-2. Specifically, how are gain and beamwidth related?

6. In steps 18 and 19, you computed the antenna gains with and without reflectors at both 1 and 2 m. How closely did your gain calculations agree with each other in both instances? Does the agreement in gain calculations at each point give you confidence that your calculated gains are close to the theoretical prediction? Why?

7. For the antenna specified in the equipment list for this experiment, the design frequency is 915 MHz. Discuss the implications for your gain measurements and calculations if the antenna is not truly resonant at 915 MHz.

FIBER OPTICS

OBJECTIVES:

    1. Become familiar with fiber-optic cabling technologies.

    2. Measure the signal loss on a fiber-optic cable of various lengths.

    3. Determine the length of fiber-optic cabling based on the optical time-domain reflectometer (OTDR) measurements.

    4. Find the location of a fault in a fiber-optic cable.

REFERENCE:

    Refer to Section 16-8 in the text.

EQUIPMENT:

    Single and multi-mode fiber optic cables in various lengths

    Optical Time Division Relectometer (OTDR) (TEKRanger TFS3031, Corning OTDR Plus Multitester II)

    Launch Box (Fibertron single mode, Fiberton multi mode)

COMPONENTS:

    ST-ST patch cable

    SC-SC patch cable

INTRODUCTION:

    Fiber-optic cable transmits light instead of voltages. There is very little signal loss in fiber-optic cables, which makes them very good options for long distance cabling needs. A copper cable can be tapped into and the information that is transmitted can be intercepted. This is much more difficult to do with a fiber-optic cable. Copper cables are affected by electromagnetic interference; however, fiber-optic cabling is not. A fiber optic cable has the following parts as shown in Figure 31-1.

    The core is the glass fiber through which the signal travels. The cladding is designed to reflect the light signal down the fiber. The buffer gives the cable strength. The insulating jacket protects the cable. Fiber optic cable is identified

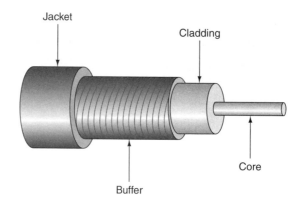

Jacket

Cladding

Core

Buffer

FIGURE 31-1   Parts of a fiber-optic cable.

with a two-number designator, such as 62.5/125 μm, that indicates the core and cladding measurements, respectively. Typical network fiber-optic cabling consists of a pair of cables: one for sending and one for receiving. This is known as a duplex fiber-optic cable.

The two common types of fiber-optic cable are single mode and multi-mode. Single mode uses a laser as the light source with a wavelength around 1310 or 1550 nm, has higher transfer rates, and can traverse longer distances. Multi-mode uses light-emitting diodes (LEDs) for the light source with a wavelength around 850 nm. Single-mode fiber-optic cable has a small core that allows only one mode of light to propagate. This limits the amount of reflections and lowers the attenuation, or signal loss, thus allowing for faster travel over longer distances. Multimode fiber-optic cable has a larger core that allows multiple modes of light to propagate. There is more attenuation because of the high signal dispersion, which reduces signal quality over long distances.

A common instrument used with fiber-optic cabling is the optical time-domain reflectometer (OTDR). An OTDR can be used to estimate the fiber length, measure attenuation, and find faults in an optical cable. The best way to measure overall attenuation in a fiber cable is to inject a known level of light in one end and to measure the light level emerging from the other end. The difference in the two levels is measured in decibels (dB) and is the end-to-end attenuation or insertion loss. An OTDR will generally show a plot indicating distance and signal level. Recall that a decibel (dB) is a unit used to express relative differences in signal strength. A decibel is expressed as the base 10 logarithm of the ratio of the power of two signals: $dB = 10\log_{10}(P_1/P_2)$, where $\log_{10}$ is the base 10 logarithm, and P_1 and P_2 are the powers to be compared. Signal amplitude can also be expressed in dB. Power is proportional to the square of the amplitude of the signal. Therefore, amplitude dB is expressed as: $dB = 20\log_{10}(V_1/V_2)$ where V_1 and V_2 are the amplitudes to be compared.

Some signal loss occurs in any fiber cable interconnection. This loss is the difference in power that you see when you insert the device into the system. For example, if you were to measure the optical power, P_1, through a length of fiber and subsequently were to cut and terminate the fiber and then measure the power again and call the output power P_2, then the difference between P_1 and P_2 would represent the connector loss. In many fiber-optic network designs it is common to assume a connection loss of 0.50 dB for each fiber-to-fiber connection.

PROCEDURE:

1. To determine the length of an unknown fiber-optic cable we will use the Corning OTDR Plus Multitester II.

2. Power on the OTDR.

3. Change the settings for the following: turn on Autosense, and change the scale to feet.

4. Connect the single mode launch box to the OTDR's SM connection.

5. Connect the unknown cable to the other end of the launch box.

6. Start the OTDR scan. You should see a screen similar to Figure 31-2.

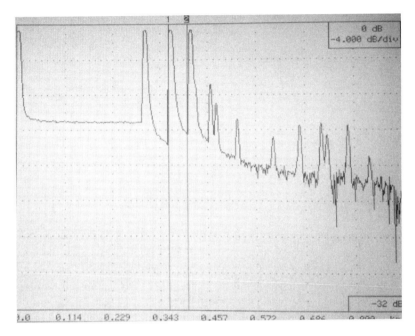

FIGURE 31-2   OTDR scan of unknown fiber-optic cable.

7. Adjust the cursor so they are at the beginning of each peak.

8. Write down the measurements between the three peaks. This is the distance between each connection as shown in Figure 31-3. The first distance is the length of the launch box.

9. Measure the distance of all the connected cables and write them down:

   a. Launch box: _____

   b. Cable 1: _____

   c. Cable 2: _____

10. You will also see the signal loss in dB for each measurement between the cursors.

11. Measure the signal loss of all connected cables and write them down:

   a. Launch box: _____

   b. Cable 1: _____

   c. Cable 2: _____

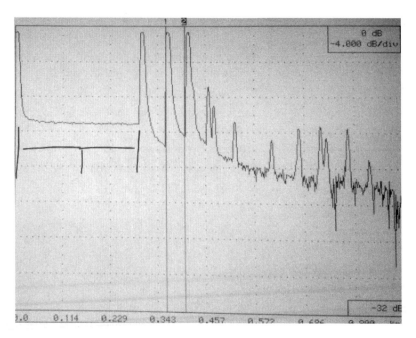

FIGURE 31-3 Measurement between peaks.

QUESTIONS:

1. If you have a 100 m run of 62.5/125 fiber that has a loss of 3.5 dB/km and connects through three patch panels (with a 0.50 dB loss per connection), what is the total loss in dB?

2. You use an OTDR on an unknown length of cable. The cable is 62.5/125 multi-mode fiber. The cable is rated for loss of 4.0 dB/km. If the OTDR shows a loss of 5.25 dB, how long is the cable?

3. Describe the major differences between single-mode and multi-mode fiber-optic cable.

4. What are some reasons to choose fiber-optic cable with respect to security features?

FIBER-OPTIC CABLE SPLICING

OBJECTIVES:

    1. To become familiar with the proper use of strippers and cleavers in the preparation of fiber-optic cable for fusion splicing or mechanical splicing.

    2. To become familiar with the proper cleaning techniques in the preparation of fiber for fusion splicing or mechanical splicing.

    3. To become familiar with fusion splicing and/or mechanical splicing techniques.

REFERENCE:

    Refer to Section 16-6 of the text.

EQUIPMENT:

    Fiber-optic inspection microscope—FIS model F1-0111-E

    Black bench mat tube—Anixter model 167438

    Safety glasses

    Alcohol bottle—FIS model F1-000728

    Fiber-optic disposal container—FIS model F1-8328

    Lint-free tissue

    Canned air—FIS model F1-1007

    Tweezers

    Jacket strippers—FIS model F1-0016

    Miller strippers—FIS model WO-1224

    Clauss strippers—FIS model WO-1225

    Clauss No-Nik strippers—FIS models NN175 and NN203

    Microstrip tool kit—FIS model MSFOK1

    Diagonal pliers

    Regular scissors

    Kevlar scissors—FIS model F1-0020

    Pocket cleaver—FIS model WO-2220

    Fitel cleaver—FIS model F1-0010

    Thomas & Betts cleaver—FIS model 92208

    Alcoa Fujikura cleaver—FIS model CT07

    Single-mode bare fiber-optic cable—FIS model SMF-28

Multimode simplex fiber-optic cable—FIS model 601-IN-SFS-62PFD/s

Paper towels

Toothpick

Cellophane tape

Stick-on labels

3M Fibrlok mechanical splice—FIS model 2529

3M Fibrlok mechanical splice assembly tool—FIS model 2501

Siecor camsplice—FIS model 95-000-04

Siecor camsplice assembly fixture—FIS model 2104040-01

AMP corelink—FIS model F04590

AMP corelink assembly fixture—Anixter model

Fusion splicer

INTRODUCTION:

The purpose of this experiment is to demonstrate fiber-optic cable-splicing techniques. Two main methods of cable splicing are in use: mechanical splicing and fusion splicing. Both techniques rely critically on the proper use of hand tools and proper cleaning techniques to ensure a high-quality splice. The processes necessary for professional-grade splices will be reviewed in detail for mechanical splices. In addition, fusion splicing techniques will be illustrated for those with access to the appropriate equipment.

SAFETY RULES:

It is mandatory that safety and cleanliness procedures be followed when working with glass fibers. Following these procedures will ensure that stray fibers do not end up in your skin and eyes. It will also ensure that stray fibers and dirt do not end up in your tools and fiber optic hardware. The following rules should be followed when working with a fiber optic installation:

1. Always work on a dark bench surface with adequate lighting so that stray fibers can always be located.

2. Always dispose of fiber scraps in a special disposable container that can be disposed of separately from other waste products.

3. Always wear safety glasses when working with fiber. Never rub your eyes when working with fiber installations.

4. Always wash your hands before and after working with a fiber installation project.

5. Check your clothing and workbench for fiber scraps after completing a fiber installation.

6. Always clean connectors and glass fibers with alcohol and lint-free tissue when preparing them for connectorization.

7. Always blow out all connectors and tool openings with canned air.

8. Keep your workbench neat and organized throughout the fiber installation project. Keep all small parts in a special area so that they will not be misplaced when you need them.

9. Never look into a fiber connector when you are not sure if it is active. The light produced by an active connector is usually invisible but could be harmful to your eyes.

10. Try not to breathe the fumes produced by the chemicals (alcohol, epoxy, solvents, etc.) used in the fiber installation process.

PROCEDURE:

1. To set up your bench for this experiment, follow these procedures:
   a. Wash your hands to ensure that all body oils have been removed.
   b. Retrieve each of the following items for your bench top. These items make up your fiber optic workbench supply kit, which is also referred to in Experiment 33.
      (1) Black bench mat tube
      (2) Safety glasses
      (3) Alcohol bottle
      (4) Fiber optic disposal container
      (5) Lint-free tissue
      (6) Canned air
      (7) Tweezers
   c. Make sure that you are working in a well-lit area. You may need to use a portable light fixture at your workbench.
   d. Open the mat and place it in the center of the work area.
   e. Put on your safety glasses and wear them whenever glass fiber scraps may be in the work area.
   f. Identify an area on your workbench where all small parts will be kept for easy retrieval.

2. Obtain a few 2-foot lengths of single-mode bare fiber optic cable and multimode simplex fiber optic cable. In the next few steps, you will be preparing these fibers for mechanical or fusion splicing. Also, obtain as many of the hand tools found in the supply list as your laboratory can offer. To view the end of your fiber under your inspection scope, the fiber should be stripped and cleaved as shown in Figure 32-1(a) so that it will fit in the recommended test fixture given in Figure 32-2. To prepare your fiber for actual splicing, it should be stripped and cleaved as shown in Figure 32-1(b). The next few steps will show you how to strip and cleave your fiber.

3. The test fixture given in Figure 32-2 should be constructed to allow for viewing of the end of the fiber with the inspection microscope. The microscope is inserted into its plastic base, which is held in place on the bottom layer of pre-punched prototyping board, using doublesided tape. Also, a splice protection sleeve (FIS model F1-1002) is secured to the five layers of pre-punched prototyping board by two $^3/_{16}$-inch cable clamps, each held in place by a small screw and locknut. Notice that the end of the protection sleeve is pushed under the slide clip of the microscope. The fiber being viewed is then carefully inserted and slid through the protection sleeve until it appears in view. Cellophane tape can then be used to hold the fiber in place. This fixture works better with the 900-micron diameter buffered

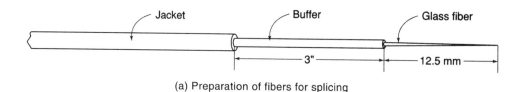

(a) Preparation of fibers for splicing

(b) View of prepared fiber on inspection scope

FIGURE 32-1 Fiber ends as viewed under inspection scope.

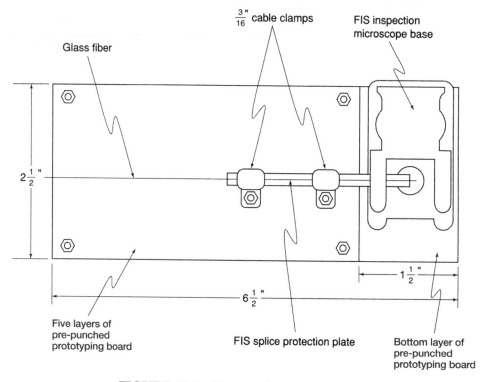

FIGURE 32-2 Construction of test fixture.

multimode fiber. If a fusion splicer is available in your laboratory, then this test fixture may not be needed because the fusion splicer may then be used to view the end of the fiber.

4. Stripping

   a. If your fiber has a jacket, it first must be stripped back. Use the jacket strippers to do this. Use the #16AWG size stripping hole in ¼-inch increments to ensure that the glass does not break inside. Strip the jacket back at least 8 inches to give you plenty of buffered glass to work with. With the multimode fiber, first tie a knot in the middle of the length of the fiber. This knot will prevent the buffered fiber from accidently slipping completely out of the jacket while you are stripping off the ¼-inch increments because you are working with only a short length of fiber. Make sure that the knot is not tied too tight or else you could break the fiber!

b. Blow on the end of your stripped cable to force all the Kevlar yarns to move to one side of the buffered fiber. Then carefully twist the Kevlar yarns into a tight strand as it comes out of the jacket to make them easy to cut. Use the special Kevlar scissors to cut the Kevlar strands so that they are completely removed at the jacket.

c. Review the manufacturer's instruction sheets for each of the Miller, Clauss, and Microstrip stripping tools. Practice stripping the buffer of the bare glass using each of the following stripping tools. Remember to strip the buffer back at least 6 inches to be able to produce the final dimensions given in Figure 32-1.

(1) Jenson red-handled jacket strippers (These will not work very well!)

(2) Miller yellow-handled buffer strippers

(3) Clauss yellow-handled jacket/buffer strippers (Use the small hole for stripping the buffer.)

(4) Clauss No-Nik mustard-handled buffer strippers (175-micron opening for 250-micron diameter buffered single-mode fiber.)

(5) Clauss No-Nik red-handled buffer strippers (203-micron opening for 900-micron diameter buffered multimode fiber.)

(6) Microstrip tool kit buffer strippers (with proper insert guide and blades)

5. Cleaning

a. Once your buffer has been stripped off, the bare fiber will most likely have small pieces of leftover fiber on it. These must be cleaned off with a lint-free tissue and alcohol.

b. Take one lint-free tissue and fold it along its crease into a square. Then fold it down the middle two more times, once horizontally and once vertically, to form a small square. Then fold it one more time down the middle to form a V-groove, such as the one shown in Figure 32-3. Hold the lint-free tissue between the index finger and thumb of your left hand. Open up the V-groove a little and drop four to five drops of alcohol into the V-groove. Then lay the stripped fiber into the V-groove and carefully pull the fiber out of the lint-free tissue while slightly pinching the lint-free tissue against the fiber.

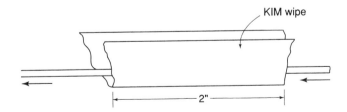

FIGURE 32-3   Illustration of cleaning procedure with lint-free wipe and alcohol.

c. Repeat this cleaning procedure with another few drops of alcohol. You should note that once the glass is clean, it squeaks just like when you are washing a window with an alcohol cleaner.

d. The lint-free tissue can be used again for other fibers after unfolding your three folds and refolding to use another part of the lint free tissue. (This keeps your bench free of used lint-free tissues.)

6. Cleaving

a. The objective here is to produce a clean cut of the end of the fiber so that the end is perfectly perpendicular to the center-line axis of the fiber, as shown in Figure 32-4. To do this, the cleaving tool first scratches the outside surface of the glass where the cut is to be made. Then a perpendicular force is firmly applied at the exact location of the scratch. The glass should then break cleanly at the desired location.

Good cleave

Poor cleave

FIGURE 32-4 Illustration of good and poor cleaves.

b. Review the manufacturer's instruction sheets for the following cleavers that are available in your laboratory. Practice the cleaving process with your fibers using each of the following tools:

(1) Diagonal pliers—will crush the glass, producing unacceptable results.

(2) Regular scissors—will crush the glass, producing unacceptable results.

(3) Pocket cleaver

(4) Fitel cleaver

(5) Thomas & Betts cleaver

(6) Alcoa Fujikura cleaver

c. Using the inspection microscope and fixture, observe the end of the fiber after using each of the cleaving tools. Carefully rotate the fiber once you have its cleaved end clearly in view. If it is cleaved properly, it should appear as the good cleave shown in Figure 32-4 when viewed at all angles. Make a sketch of your findings for each of the cleavers used in step 6b. You should find that the more expensive cleavers will produce acceptable results more consistently.

d. It is absolutely necessary that an acceptable cleave be produced before any splicing can be successful. An unacceptable cleave will not allow the two fibers to be spliced and to be butted next to each other in perfect alignment. Thus, the splice will end up with a high loss. Therefore, do not proceed to step 7 or 8 until acceptable cleaves have been produced.

e. The fibers being prepared for splicing in either step 7 or 8 must have their other ends terminated with an appropriate connector that can be connected to a light source and power meter or to an optical time domain reflectometer (OTDR). This termination is necessary for measuring the resulting power loss of the splice. For the use of an OTDR, it is recommended that two 100- to 150-meter rolls of fiber be used in steps 7 and 8, as shown in Figure 32-5. Most OTDRs cannot make accurate measurements on the first 10-80 meters of fiber that are connected to their laser output connectors due to saturation of their detectors by the high amounts of scattered light energy. This is referred to as the OTDR's dead zone.

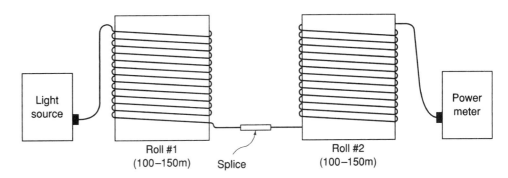

FIGURE 32-5   Use of OTDR to measure lengths of fiber-optic cable beyond OTDR dead zone.

7. Mechanical Splicing (Optional)

a. A mechanical splice is a small plastic device in which prepared fiber ends are inserted from either end into V-grooves. The center of the V-groove, where the ends butt together, is filled with index matching gel to help create the lowest loss possible.

b. Review the manufacturer's instructions for the mechanical splices listed below that are available in your laboratory. Note that each of the mechanical splices requires the use of a special test fixture to assemble a low-loss splice properly. Some of the splices can be attempted repeatedly until acceptable results are achieved. This may require the use of special small hand tools to reopen the mechanical splice. Eventually, however, the index matching gel inside the V-groove will become contaminated with dirt and broken pieces of fiber. Then acceptable results can never be achieved. At that point, repreparation of the fibers and a new mechanical splice will be necessary.

(1) 3M Fibrlok

(2) Siecor Camsplice

(3) AMP Corelink

c. To minimize the power loss in the mechanical splice, a light source is placed at one end of the fiber and the power meter is placed at the other end. The fibers are adjusted in the mechanical slice until minimal loss is achieved. (The procedure for measuring actual loss is covered in more detail in Experiment 33.)

d. You should notice that the more expensive mechanical splices are easier to use. They make more consistent low-loss splices and are more capable of being readjusted for optimum results.

8. Fusion Splicing (Optional)

   a. A fusion splice is created by heating the glass fibers with a predetermined arc of electrical energy. As the glass melts, the fibers are carefully pushed together in near-perfect alignment so that, when the glass solidifies, it becomes a continuous piece of fiber.

   b. If your laboratory is equipped with a fusion splicer, review the manufacturer's instructions for its use. The earlier models require the operator to align carefully the two fibers being fused by manual adjustments along both horizontal and vertical axes of view. The later models perform this alignment automatically by computer.

   c. Practice these procedures with your cleaved fibers until you feel that you have mastered these skills. Again, to check your resulting splice loss, a light source and power meter are attached at opposite ends of the fiber being tested and measurements are made. The alternate procedure is to "shoot" one end of the fiber with an OTDR laser source and observe the resulting pattern on the OTDR display.

9. Clean-up—It is important to follow these procedures for your protection and to prolong the functional life of your tools and fiber installation project.

   a. Cover all fiber optic connectors that are not in use with their safety caps.

   b. Clean all tools that may have been exposed to pieces of glass fiber and stripped buffer. Use canned air, piano wire, cleaning brushes, lint-free tissue, and alcohol to remove all scraps, dirt, and debris from all tools and your work area.

   c. Carefully inspect the workbench and your clothing for broken fibers. Pick up any stray fibers with tweezers and drop them into a disposable container.

   d. Return all tools and working mat back to their original containers and store them in their proper location.

   e. Wash your hands and face with soap and water.

QUESTIONS:

1. Why is it so important to follow safety and cleanliness procedures when working with fiber optic cable? List the most important procedures that must always be followed.

2. Why did you need to tie a small knot in the 2-foot fiber optic cable before stripping off the jacket?

3. Which of the buffer stripping tools worked best for you? List any problems you have with any of the buffer strippers.

4. What is the advantage of folding the lint-free tissue as detailed in step 5?

5. Which of the cleavers produced ideal results most consistently when viewed under the fiber inspection microscope? List any problems you had with any of the cleavers listed in step 6.

6. Which of the mechanical splices in step 7 seemed to produce low loss most consistently? Which was the easiest to readjust for optimum results?

FIBER-OPTIC CABLE CONNECTORIZATION

OBJECTIVES:

1. To become familiar with the procedure used to connectorize multi-mode fiber-optic cable with ST connectors.

2. To become familiar with the procedure used to measure the loss of connectors on a patch cord using a light source and power meter.

3. To recognize the differences between ST connectors that have stainless steel ferrules and those that have ceramic ferrules.

REFERENCE:

Refer to Section 16-6 in the text.

EQUIPMENT:

Fiber-optic light source—FIS model 9050-0000

Fiber-optic power meter—FIS model F1-8513HH with universal adapter

Fiber-optic inspection microscope—FIS model F1-0111-E

Stainless-steel ferrule connectors—FIS model F1-0066 (2)

Ceramic ferrule connectors—FIS model F1-0061 (2)

Four-foot piece of simplex multi-mode fiber-optic cable—FIS model 601-IN-SFS-62PFD/S

ST mating sleeve—FIS model F1-8102

Fiber-optic workbench supply kit (detailed in Experiment 32)

Fiber-optic jacket strippers—FIS model F1-0016

Fiber-optic buffer strippers—any of those detailed in Experiment 32

Kevlar scissors—FIS model F1-0020

Scribe—FIS model F0-90C

Piano wire—FIS model F1-8265

Epoxy—FIS model F1-7070

Glass polishing plate—FIS model F1-9111A

Soft pad—FIS model 05-00053

Polishing disk—FIS model F1-6928

Polishing paper—FIS models F1-0109-5, F1-0109-1, and F1-0109-3

Connector heat oven (optional)—FIS model F1-9772 with 9451S curing stand

Paper towels

Toothpick

Cellophane tape

Stick-on labels

INTRODUCTION:

This experiment will introduce you to the steps for placing ST-type connectors ("connectorizing") on multimode fiber optic cable. In addition, after installing the connectors, you will measure signal attenuation to verify the integrity of your completed assembly.

PROCEDURE:

1. The consumable supplies for this experiment are the ST connectors, multimode simplex fiber optic cable, and polishing paper. The ST connectors consist of three parts: the back shell, the rubber boot, and the safety cap, as shown in Figure 33-1. Two of the ST connectors have a stainless steel ferrule (silver) and two have a ceramic ferrule (white). The ceramic ferrule connector is less forgiving in the connection procedures because the ceramic material will not polish like the stainless steel will. Carefully store these parts in a safe location because they are small enough to get lost easily. There are three types of polishing paper: coarse (5-μm grit) black paper, medium (1-μm grit) green paper, and fine (0.3-μm) white paper.

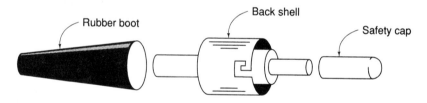

FIGURE 33-1   ST connector.

2. **Setup**—Set up your lab bench following the procedures outlined in Experiment 32. You will also need the following tools at your lab bench:
   a. Fiber jacket strippers
   b. Fiber buffer strippers
   c. Kevlar scissors
   d. Scribe
   e. Epoxy—Tube A and Tube B
   f. Fiber optic crimping tool
   g. Glass polishing plate
   h. Soft pad

    i. Polishing disk (puck)

    j. Paper towel

    k. Toothpick

    l. Cellophane tape

    m. Small stick-on labels

3. ***Initial Preparation***—To prepare for two patch cords, you need to cut your 3- to 4-ft fiber optic cable into two equal lengths. Cut your cable in the middle of its length with the cutting blade of your jacket strippers. This will cut the jacket and the glass fiber. Then you will need to cut through the Kevlar yarns with the special Kevlar scissors. Next, tie a knot in the middle of each of your two patch cords. Do not make the knot so tight that you break the fiber. It does have to be tight enough to prevent jacket slippage when you strip the buffer in step 4. Also, place sticker labels on each of your patch cords. Label with your last name followed by SS or CM, for stainless steel or ceramic ferrules, respectively. This labeling will prevent your patch cord from being mixed up with another student's cable when it is placed in the oven or being tested for loss with the power meter.

4. ***Stripping***—The ends of each of the two patch cords need to be stripped back so that they match Figure 33-2. To produce these results, the following procedure should be followed for each of the four cable ends:

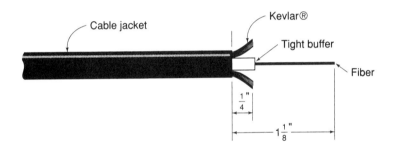

FIGURE 33-2   Illustration of patch cord end with insulating layer stripped away.

    a. Strip back the jacket approximately $1\frac{1}{8}$ inch to expose the buffer and Kevlar yarn. Use your jacket strippers. Use the #16AWG size stripper hole and strip in $\frac{1}{4}$-inch increments to ensure that the glass does not break inside.

    b. Blow on the end of your stripped fiber to force all the Kevlar yarns to move to one side of the buffered fiber. Then carefully twist the Kevlar yarns into a tight strand as they come out of the jacket to make them easy to cut. Use the special Kevlar scissors to cut the Kevlar strands so that they all stick out away from the jacket approximately $\frac{1}{4}$ inch.

    c. Next, use the buffer stripping tool to strip the buffer away from the glass. The glass (core and cladding) should stick out approximately $\frac{7}{8}$ inch beyond the end of the buffer to conform to the dimensions given in Figure 33-2. Stripping the buffer should be done in short increments if you are using a Clauss or Miller hand stripper. If you are using the No-Nik or Microstrip strippers, you should be able to strip $\frac{5}{8}$ inch in one complete pass. Don't forget to clean your stripping tool with canned air when you are finished stripping the fiber.

5. *Cleaning*—Carefully clean the glass with alcohol and lint-free tissue. Refer to step 5 of Experiment 32 for detailed instructions on the cleaning procedures.

6. *Epoxying/Crimping*—Two patch cables will be fabricated. The first one will use the stainless-steel ferrule connectors and the second will use the ceramic ferrule connectors. The following procedure should be used for the stainless-steel ferrule connectors:

   a. Carefully slide on the rubber boot. Make sure that the wide end of the boot faces the back end of the back shell. It must be installed now because it won't slide over your connector later!

   b. Look through the back end of the back shell of the ST connector using an eye loupe and back light to make sure that there is no dirt or obstruction inside the passageway that would interfere with the glass fiber's ability to be pushed through the connector. If there is an obstruction, blow it out with canned air or push it out with piano wire.

   c. Try a "dry run." Carefully feed your glass fiber into the back end of the back shell of the ST connector and see if it will slide out of the tip of the ferrule. Be very careful not to trap the fiber and subsequently break it by making it bend against itself. This procedure is very delicate like threading a needle, and practice is necessary. You should notice that you will be able to continue feeding the glass fiber through the connector until the buffer jams into the back end of the ferrule. It will not be narrow enough to enter the ferrule opening. Once you feel that you have mastered this procedure, continue with the next step.

   d. Now it is time for the real thing! Mix a small blob of epoxy by using a 2:1 ratio of epoxy "A" and epoxy "B." Do this on a scrap of paper or paper towel. Stir the two quantities together with a toothpick or equivalent.

   e. Using the toothpick, carefully paint the glass fiber, buffer, and Kevlar yarn with a thin coating of mixed epoxy. Then carefully feed your epoxy coated glass into the back of the main shell just as you practiced in step 6c (see Figure 33-3). If you are right-handed, hold the connector with your left hand and guide the fiber into the connector with your right hand. Continue pushing until the buffer hits the back entrance of the ferrule. The glass fiber should protrude out of the tip of the ferrule. Be very careful not to let the glass break as it protrudes out of the opening of the ferrule. If it does break, you will have to start the whole preparation process again! If you are left-handed, hold the connector with your right hand and guide the glass fiber in with your left hand.

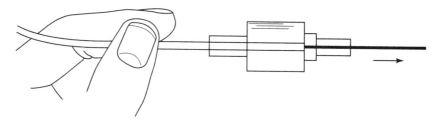

FIGURE 33-3 Placement of main shell onto epoxy-coated glass.

f. As the jacket gets closer to the back of the connector, the epoxy covered Kevlar should fold back against the jacket and tuck in between the jacket and the side of the back shell of the connector. This will allow for sealing the fiber cable into the connector after the crimp is made and the glue has hardened. With the fiber fully inserted into the back of the connector, very carefully position the side of the back shell against the .128 die of the crimping tool. Start closing the crimping tool (see Figure 33-4). Once the crimp is started, it is impossible for the crimpers to be reopened until the crimpers have been fully closed, completing the crimp. You will need to push hard with both hands to complete the crimp. While you are pushing hard with the crimpers, be careful to position the front of the connector so that the protruding glass fiber does not hit something and break off.

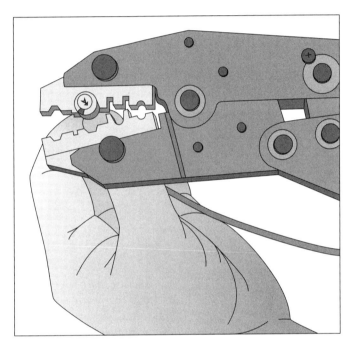

FIGURE 33-4   Illustration of crimping connector with fiber-optic crimping tool.

g. Carefully remove the connector assembly from the die of the crimpers without breaking the glass. Now the epoxy must be left to harden, which takes twenty-four hours at room temperature or thirty minutes at 100° C. If you are using an oven to speed the curing process, carefully guide the connector into the circular opening of the oven so you do not break off the fiber strand that is sticking out of the tip of the ferrule. Using cellophane tape, secure the jacket of the fiber to the side of the curing stand. Don't forget to clean any leftover epoxy from the die of the crimping tool.

h. Repeat steps 6a through 6g to place a connector on the opposite end of the patch cord.

i. For the second patch cord, which is made of the ceramic ferrule connectors, the procedure is the same except that, after completing step 6f, place a small bead of epoxy on the tip of the ferrule, around the protruding glass fiber. This ensures that the fiber does not accidentally break below the ferrule tip surface during the polishing process. The ceramic material will not polish like the stainless-steel material does. The cured epoxy will polish away.

7. **Scribing**—Carefully remove the crimped connector assembly from the oven or storage area when your epoxy has completely cured. Be very careful at this point because if the glass fiber strand is accidentally bumped, it could easily break off, making it difficult if not impossible to polish. Be aware that the back shell may need five minutes to cool enough so that you can handle it. Carefully slide the rubber boot over the crimped part of the connector to give the fiber more strain relief where it comes out of the back of the connector. Now complete the scribing procedure as outlined in Figure 33-5.

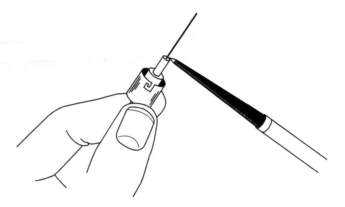

FIGURE 33-5   Illustration of scribing procedure.

   a. Using a scribe, carefully scratch the side of the glass fiber close to but just above the ferrule tip surface. One scratch should be sufficient. If you scratch several times, the glass will most likely break off in an unpredictable location.

   b. Next, carefully pinch the end of the glass fiber with your thumb and index finger and pull the glass away from the tip of the ferrule. The glass strand should break away fairly easily at the exact location of the scratch. This should leave you with a small nub of glass sticking out of the end of the connector. Dispose of the glass fiber scrap properly. Make sure that the nub is not too long. If it is too long, the next step of air polishing may take an excessive amount of time or result with a broken glass nub. Thus, you may want to try to rescribe a bit closer to the ferrule tip surface if it is possible.

8. **Air Polishing**—The objective of this procedure is to polish away the glass nub so that the glass fiber ends right at the ferrule tip surface. Refer to Figure 33-6 and complete the following steps:

   a. Take a sheet of black polishing paper (5-μm grit) and hold it at one end so that the other end is hanging in free space with the coarse (dull) side facing up. You need to bend the polishing paper in a slight U-shape to give it a little stress support.

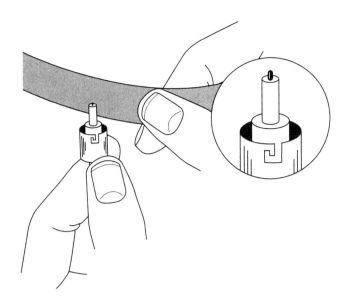

FIGURE 33-6 Polishing of glass end of fiber-optic cable.

    b. Push the glass nub down very lightly against the opposite end of the polishing paper on its rough side. Begin rubbing the tip of the nub against the polishing paper in a small circular motion. Carefully repeat this polishing action without pushing too hard. You should notice that a fine circular scratch is being drawn on the polishing paper.

    c. After a dozen circular traces have been completed, view the end of the ferrule using your eye loupe. You should notice that the length of the nub has decreased due to the polishing action.

    d. Carefully continue this polishing process until the nub has been completely polished away. When this happens, if you are using the stainless-steel ferrule connectors, you will find that the circular traces on the polishing paper will take on a much thicker appearance because you are now starting to polish the stainless-steel surface along with the end of the fiber. If you are using the ceramic ferrule connectors, you will notice that the scratching sound will disappear and you will not be polishing anything anymore because the ceramic ferrule cannot be polished by the polishing paper.

    e. Be careful that you don't push too hard or the nub will break off unpredictably and you may have to start the entire procedure again!

    f. When the nub has been completely polished away, you are now past the critical point in the connection process. The connector assembly is now not as delicate and prone to breakage due to mishandling. You can now proceed to the polishing procedure.

9. *Polishing* — Now the objective is to make the end of the glass fiber at the tip of the ferrule as smooth and clear as possible. This mirror surface is achieved by the use of three polishing papers, a glass polishing table, and a polish disk (puck). The following procedures are followed with the stainless-steel ferrule connectors:

    a. Using the fiber inspection microscope, view the tip of the ferrule. It should look like a fuzzy black dot on a scratchy silver or white background. Check for any evidence of broken glass pieces. It is hard to determine if the fiber is broken at this time.

b. Place the black (0.5-μm grit) polishing paper on the glass polishing table so that the rough side faces up. Clean the polishing puck and place it on top of the polishing paper. Now insert the ferrule of the connector into the polishing puck and push it down until it cannot be inserted any further. Lightly push the puck/connector assembly around in a figure-eight pattern so that all edges are polished equally. Approximately eight to ten figure-eight strokes should suffice.

c. Remove the connector from the puck. Clean the end of the ferrule with alcohol and a lint-free tissue. When it is squeaky clean, let the alcohol evaporate and then view the tip of the ferrule with the inspection microscope. It should now look more like a gray dot against a silver or white background. If not, repeat step 9b until it does. There may be a few scratches and epoxy spots on the glass at this point. The use of finer grit polishing paper should take care of this. What you do not want to see is a dark black dot representing the glass surface. If this happens, you most likely have broken the fiber inside the ferrule and you will have to start the whole process again!

d. Clean the bottom of the polishing puck with alcohol and a lint-free tissue. This keeps the new polishing paper from being contaminated with the scraps from the old polishing paper. Now use the green polishing paper (1-μm grit) and repeat step 9b. This should remove more of the remaining scratches and epoxy spots from the surface of the glass fiber. Again, eight to ten figure-eight strokes should suffice.

e. Again, remove the connector from the polishing puck. Clean the end of the ferrule with alcohol and a lint-free tissue. You should again hear the ferrule squeak clean when it has been totally cleaned off. Once the alcohol has evaporated, view the tip of the ferrule with the inspection microscope. It should have less scratching and spotting, and the glass should be a light gray color. If not, repeat step 9d until it does.

f. Clean the bottom of the polishing puck with alcohol and a lint-free tissue to avoid contaminating the new polishing paper. Now use the white polishing paper (0.3-μm grit) and repeat step 9b. This should remove most of the remaining scratches and epoxy spots from the surface of the glass fiber. This time, you should complete twenty to twenty-five figure-eight strokes.

g. Again, remove the connector from the puck. Clean the end of the ferrule with alcohol and a lint-free tissue until it is squeaky clean. Once the alcohol has evaporated, view the tip of the ferrule with the inspection microscope. It should be almost completely free of scratches and epoxy spots. The glass should be a clear, light gray circle. If not, then complete steps 9h and 9i. If it is clear, then skip steps 9h and 9i and proceed to step 9j.

h. Clean the bottom of the polishing puck with alcohol and a lint-free tissue. This time, place a few drops of alcohol on the white polishing paper in an unused area. Continue polishing the connector by pushing the puck/connector assembly through the alcohol drops as they evaporate from the surface of the polishing paper. This should remove the last few specks of epoxy from the glass surface.

i. Make one last inspection of the ferrule with the inspection microscope. It should now look perfectly clear of epoxy specks and scratches. You do not want to see any of the last five displays shown in Figure 33-7. If you do, repeat step 9h until the ferrule is perfectly clear.

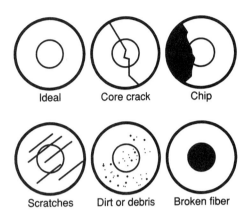

FIGURE 33-7 Ideal and defective splices as seen under inspection microscope.

j. Obviously, steps 9a through 9i must be completed for the opposite end of the patch cable.

k. For the patch cable using the ceramic ferrule connectors, the polishing procedure is identical, except that you should place a soft pad (sheet of rubber) between the polishing paper and glass. Also steps 9d and 9f may take more than twenty to twenty-five figure-eight strokes. The soft pad allows you to remove extra epoxy deposits that may become trapped on the glass below the surface of the ferrule tip. Remember that it is impossible for the polishing paper to polish away any of the ceramic material.

10. *Protection*—To protect the ferrules from dirt and scratches, make sure that you install the protective caps on the ferrules when they are not in use.

11. *Connector Loss Measurement*—To determine the loss of each of the connectors at either end of the fabricated patch cord, the following procedure must be completed. Make sure that all connectors on the test equipment and the patch cords are thoroughly cleaned before any measurements are made. Also, the fiber optic cables must be kept as straight as possible to produce the smallest loss. Follow this procedure for each connector of both of the patch cords that were fabricated in steps 1 to 10. Enter the resulting measurements in Table 33-1.

a. Place a low-loss reference patch cord between the LED light source and power meter, as shown in Figure 33-8(a). The end of the patch cord nearest the light source should be wound five to six times around a mandrel to ensure that the higher ordered modes of light are not propagated through the fiber, thus causing erroneous readings to be made by the power meter. The mandrel is nothing more than a pencil or pen or wooden doweling of the same diameter. You may need to hold the fibers in place with cellophane tape.

TABLE 33-1 Connector Loss Measurement.

PATCH CABLE	CONNECTOR	REFERENCE POWER LEVEL	POWER LEVEL	dB LOSS
Stainless-steel	A			
ferrule	B			
Ceramic	A			
ferrule	B			

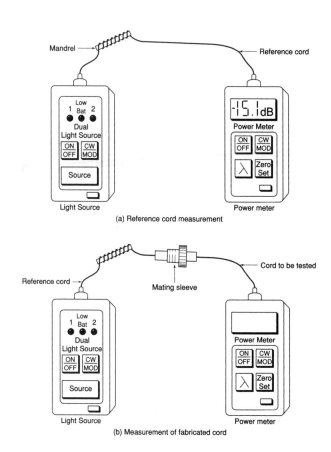

FIGURE 33-8 Illustration of power-loss measurements: (a) of reference cord; (b) of fabricated cord.

b. Turn on the light source and power meter. Make sure that the light source and power meter are designed for the same wavelength (850 nm). Measure the power level displayed by the power meter in dBm. It should be approximately –17 dBm. Turn off the light source before proceeding.

c. Disconnect the power meter from the sample patch cord. Add an ST mating sleeve and add your constructed patch cord between the reference patch cord and the power meter, as shown in Figure 33-8(b).

d. Turn on the light source. Again, measure the power level displayed by the power meter in dBm. The difference between this reading and the reading made in step 11b is the loss of the connector of your patch cord that is connected to the mating sleeve. The other connector that is plugged into the power meter is not being tested because the power meter's mating connector allows for almost all of the light to shine on the detector surface, even if the connector has substantial loss.

   e. Reverse your constructed patch cable end for end to measure the loss of the other connector of your cable. If your results in steps 11c and 11d are less than 0.5 dB, your patch cable is acceptable. If the loss is larger than 0.5 dB, you may need to reconnect one or both of your connectors!

12. Clean your workbench following the procedures outlined in Experiment 32.

QUESTIONS:

1. State the differences in preparing a fiber-optic cable for connection in comparison to preparing a fiber-optic cable for splicing.

2. Why is it so important to clean your tools, connectors, and work area constantly when fabricating patch cords?

3. Which steps did you find had the greatest chance of breaking a glass fiber? What extra precautionary steps could be taken to avoid breaking the glass?

4. List the differences in the connection of stainless-steel ferrule connectors versus ceramic ferrule connectors? How do you account for the differences?

5. What is a mandrel used for?

6. Which connector is being tested by the test procedures followed in step 11? Why aren't the other connectors shown in Figure 33-8(b) being tested?

FIBER-OPTICS COMMUNICATION LINK

OBJECTIVES:

1. To become familiar with fiber optic mounting and fabrication procedures.

2. To build and test a fiber-optic transmitter and receiver.

3. To evaluate three types of fiber-optic emitters.

4. To evaluate two types of fiber-optic detectors.

REFERENCE:

Refer to Section 16-7 of the text.

EQUIPMENT:

Dual-trace oscilloscope

Function generator

Low-voltage power supply (2)

Hot cutting knife

COMPONENTS:

Integrated circuits: LM5534 (2), LM386-3 (2)

Resistors: 10 Ω (2), 12 Ω (2), 27 Ω, 100 $\Omega-1$ W (2), 1 kΩ, 2.2 kΩ, 10 kΩ, 100 kΩ, 220 kΩ

Potentiometer (10-turn trim): 1 kΩ (2), 20 kΩ

Capacitors: 500 pF, 0.001 μF, 0.01 μF, 0.1 μF (4), 10 μF (4), 470 μF (4)

Microphone: Panasonic P-9930 miniature cartridge

Speaker: 8 Ω

Fiber-optic emitters: Motorola MLED-71, MFOE-71, low-profile red LED (e.g., HLMP-3200)

Fiber-optic detectors: Motorola MRD-721, MFOD-71 (2)

Fiber-optic cable (plastic): three 8-ft lengths of AMP 501232

Fiber-optic plugs: AMP 228087-1 (3)

Fiber-optic device mounts: AMP 228040-1, AMP 228709-1 (2)

INTRODUCTION:

In this experiment you will investigate a complete fiber-optic communications system of your own making. First, you will place connectors on a length of fiber-optic cable. Then, you will construct a transmitter and receiver capable of modulating and demodulating analog intelligence placed onto fiber-optic emitters.

PROCEDURE:

Part I: Cable Fabrication

Prepare three 8-ft fiber-optic cables with terminations as shown in Figure 34-1.

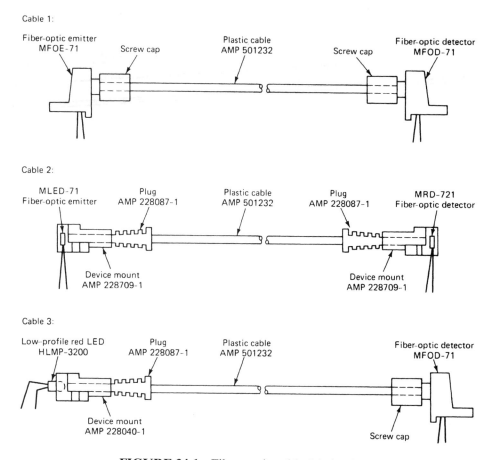

FIGURE 34-1 Fiber optic cable fabrication.

For minimal attenuation between transmitter and receiver it is essential that each of the cable ends be severed with a smooth surface. Thus a simple cut with a pair of diagonal cutters is undesired. A recommended procedure to follow is to use a hot cutting knife. A plastic cutting knife such as the Weller SP23 HK performs this task quite well. If this is not available, simply heat up an X-Acto knife with a lighted match. Apply a straight perpendicular cut with a firm downward force. Do not allow excessive burning or melting of the dark coating of the cable to occur.

To fabricate cable 1, first remove the caps of the MFOD-71 and MFOE-71. Cut the cable at both ends as described above. Mount the screw caps onto the cable. Insert the cable into the mating mount as far as it will go. Then screw the caps into the mount. This should apply a sufficient press fit onto the cable as the caps are hand-tightened.

To fabricate cable 2, the cable is cut at both ends as described above. The black jacket of the plastic cable must be stripped away from the last ½ in. of the cable at both ends. When using wire strippers to do this, be careful not to score the outer surface of the clear plastic fiber in the center of the cable. Now, insert the cable into the plugs. Needle-nosed pliers should be used to force the cable into the plugs. The ends of the cable should line up at the edge of the plug at approximately the same position where the cable cannot be inserted any farther into the plug. Now insert the plugs at each end of the cable into their respective mating device mounts. At the transmitter end of the cable, insert the MLED-71 emitter into the rectangular opening at the base of the device mount. Make sure that the LED is pointed toward the cable (white backing away from the cable). At the receiver end of the cable, insert the MRD-721 detector into the rectangular opening at the base of the device mount. Again, make sure that the detector is pointed toward the cable (blue backing away from the cable).

To fabricate cable 3, at the transmitter end of the cable follow the same procedure as in cable 2 for mounting the cable to the plug and mounting the plug to the device mount. The device mount is a different part number since it has a TO-92 opening rather than a rectangular opening for the LED. Carefully glue the LED into the TO-92 opening. Use 5-minute epoxy. Do this by carefully applying a small amount of epoxy to the sides of the LED, being careful not to smudge the top surface of the LED's lens. Insert the LED into the TO-92 opening. Let the bottom surface of the LED stick out approximately 0.1 in. away from the bottom surface of the device mount. This will ensure that the plug does not push into the top of the LED when inserted into the mount. At the receiver end of the cable, follow the same procedure that was used in cable 1 for mounting the detector to the cable.

Part II: Construction of Fiber-Optic Transmitter and Receiver

1. Build the fiber-optic transmitter given in Figure 34-2. Use the MFOE-71 of cable 1 as the fiber-optic emitter. Make sure that C_2 is placed as close to pins 4 and 6 of the 386-3 as possible. This will make sure that the amplifier does not break into oscillations. Apply ± 10 V dc to the circuit. Adjust R_6 for approximately 100 mA of current through the MFOE-71.

2. Connect a jumper between TP_1 and TP_2. Apply a 60-mV$_{p-p}$, 5-kHz sine wave as V_{in}. Adjust R_2 to maximize the amplitude of the voltage at TP_3. Measure the voltage at TP_3. Calculate the decibel voltage gain of the 386-3 transmitter amplifier using

$$A_{v(dB)} = 20 \log \frac{V_{TP3}}{V_{TP1}}$$

3. Determine the 3-dB bandwidth of the amplifier stage by finding the frequencies at which V_{TP3} drops 3 dB below its maximum value. The bandwidth should be quite large. Do not disassemble this circuit before proceeding.

4. Build the fiber optic receiver given in Figure 34-3. Again, make sure that capacitors C_7, C_9, C_{12}, and C_{13} are placed as close to the integrated circuits as possible so that oscillations will not occur. Use the MFOD-71 as the fiber optic detector. Leave TP_6 disconnected from TP_7.

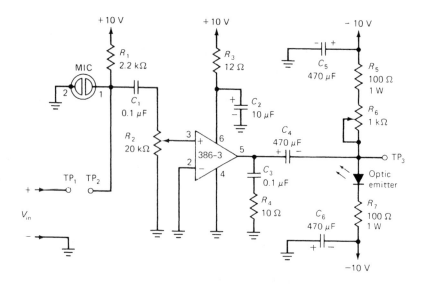

FIGURE 34-2 Fiber optic transmitter.*

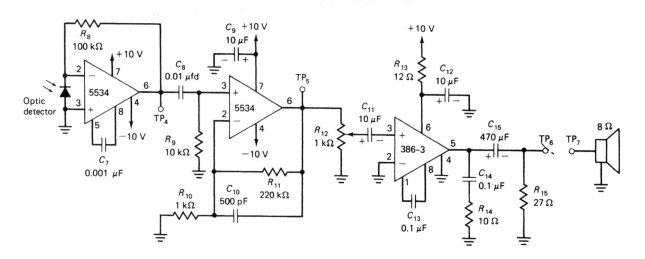

FIGURE 34-3 Fiber-optic receiver.*

5. The transmitter and receiver stages should be linked together with cable 1. Apply a 60-mV$_{p-p}$, 5-kHz sine wave as V_{in} to the transmitter. Measure V_{TP6}. Adjust R_{12} so as to force V_{TP6} to be exactly 6 V$_{p-p}$.

6. Record V_{TP1}, V_{TP3}, V_{TP4}, V_{TP5}, and V_{TP6}.

7. Repeat steps 5 and 6 with cable 1 replaced with cable 2. (Don't readjust for 6 V though.)

8. Repeat steps 5 and 6 with cable 2 replaced with cable 3. Before reenergizing the circuit, adjust R_6 to its maximum resistance. Energize the circuit and readjust R_6 for approximately 20 mA of current through the LED. Notice

*Original design and testing of these fiber-optic circuits were made by William J. Mooney and Scott Smith of the Optical Engineering Technology Department at Monroe Community College.

that the LED is giving off visible light, whereas the emitters used with cables 1 and 2 work mostly with infrared light, which is not visible. After making these measurements, temporarily remove the screw cap from the receiver end of the cable. You should be able to see visible light at the receiver end of the cable. Restore the screw cap on the end of the cable before proceeding.

9. Replace cable 3 with cable 1. Readjust R_6 for 100 mA of current flow. Disconnect the function generator from the transmitter by disconnecting the jumper between TP_1 and TP_2. Repeat steps 5 and 6. Connect a jumper between TP_6 and TP_7 of the receiver in order to connect the speaker at the output. Whistle into the microphone and you should be able to hear the whistle in the speaker. Adjust R_{12} so that the output signal is at a comfortable listening level. Too large an output voltage may cause feedback to occur between the speaker and the microphone, resulting in an undesirable squeal produced by the speaker.

10. Try each of the three fiber-optic cables. Determine by listening to the speaker which of the cables offers minimal attenuation between transmitter and receiver. Do not forget that R_6 needs to be increased to limit the current to 20 mA in the LED of cable 3.

OPTIONAL PROJECTS:

1. Determine how long each of the three cables can be made before there is so much attenuation between the transmitter and receiver that the output signal at the speaker is too small to be heard very well.

2. Place this fiber-optic link between the modulator and demodulator stages used in the digital communication systems investigated in:

   a. Experiments 13 and 14: Pulse-amplitude modulation (PAM) and time-division multiplexing (TDM)

   b. Experiment 15: Pulse-width modulation (PWM) and detection

   c. Experiment 18: Digital communication using frequency-shift keying (FSK)

   Determine how well the total digital communication system works with the fiber-optics added. List any modifications that are needed to produce satisfactory results.

QUESTIONS:

1. Determine the decibel voltage gain between the output of the receiver at TP_6 and the input of the transmitter at TP_1 for each of the three cables used in steps 5–8.

2. Based on the calculated results in question 1 and observed results in steps 9 and 10, which of the fiber optic cables produced the minimal amount of attenuation? Which produced the largest attenuation? Since each of the cables are of the same type and length, what must be the main reason for differences in the amount of attenuation? Refer to the data sheets of the fiber-optic emitters and detectors for clues.

3. Give a brief description of how the applied signal at TP_1 is processed through the fiber optic link by each stage as it proceeds to the speaker at TP_7.

APPENDIX: ARDUINO PROGRAMMING CODE AND BREADBOARD LAYOUTS FOR EXPERIMENTS 14, 16, 17, AND 18

EXPERIMENT 14: TIME-DIVISION MULTIPLEXING (TDM)

Breadboard layout for step 1:

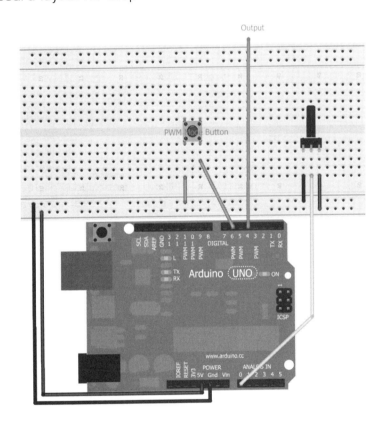

Arduino program:

```
#define PPMout 4
#define singlePWMmode 6
int channel[6]; //Channel values
int sensorValue = 5;
```

```
void pulseOut(byte pin, int timeMs)
{
    if (timeMs > 0){

      digitalWrite(pin, HIGH);
      delayMicroseconds(timeMs);
      digitalWrite (pin, LOW);
    }
}


void setup()
{
    Serial.begin(57600); //Initialize Serial/
    pinMode(PPMout, OUTPUT); // Pin 4 as
    pinMode(singlePWMmode,INPUT_PULLUP);
    for (int i=0; i<8; i++)
    {
        channel[i]= 500;
    }
}


void loop()
{

if (digitalRead(singlePWMmode)) //Is the switch triggered to
run TDM
  {
     sensorValue = analogRead(0);
     pulseOut(PPMout,sensorValue+500);
     delayMicroseconds(1024-sensorValue);
  }
else //Run TDM mode
  {
      for (int i=0; i<6; i++)
      {
        channel[i] = analogRead(i);
      }
      Serial.println(channel[0]);
      dispPWMTDM();
  }
}


void dispPWMTDM()
{
  for (int i=0; i<6; i++)
  {
    pulseOut(PPMout, channel[i]+200);
    //Serial.println(500*(n & 1));
    delayMicroseconds(1024-channel[i]);
  }
  digitalWrite(PPMout,LOW);
  delay(8);
}
```

Breadboard layout for step 1:

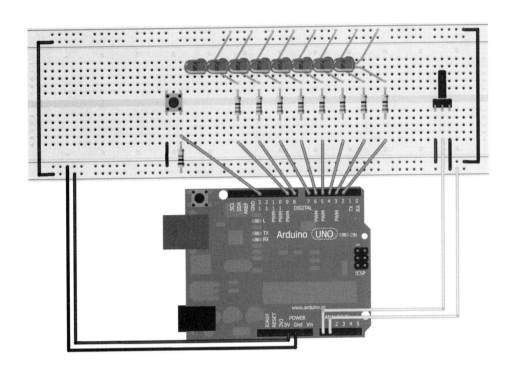

Arduino program:

```
int ledPins[] = {2, 3, 4, 5, 6, 7, 8, 9, 10}; //least-significant
to most-significant bit
byte count;
#define nBits sizeof(ledPins)/sizeof(ledPins[0])
#define SENSORPIN A0
#define DELAYPIN A1
#define CLOCK 13
int sensorValue = 0;

void setup(void)
{
  for (byte i=0; i<nBits; i++) {
    pinMode(ledPins[i], OUTPUT);
  }
  pinMode(CLOCK,INPUT);
  pinMode(SENSORPIN,INPUT);
}

void loop(void)
{
  if (digitalRead(CLOCK)==LOW)
```

```
  {
    sensorValue = analogRead(SENSORPIN);
    sensorValue = map(sensorValue, 0, 1023, 0, 255);
    dispBinary(sensorValue);
    delayMicroseconds(10*analogRead(DELAYPIN));
  }
}

void dispBinary(int n)
{
  for (int i=0; i<nBits; i++) {
    digitalWrite(ledPins[i], n & 1);
    n /= 2;
  }
}
```

EXPERIMENT 17: PULSE-CODE MODULATION (PCM) AND SERIAL DATA PROTOCOLS

Breadboard layout (step 1):

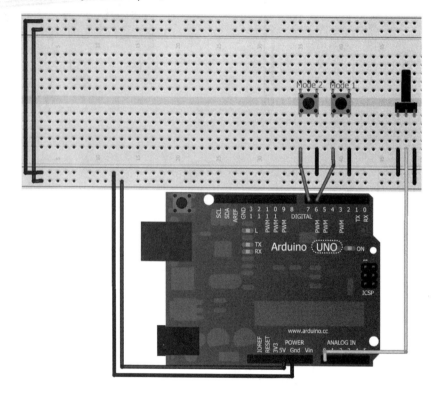

Breadboard layout (step 15):

Arduino program:

```
#include <SPI.h>
#define bitwidth 12 //Set depending on MCP4901, MCP4911,
MCP4921
```

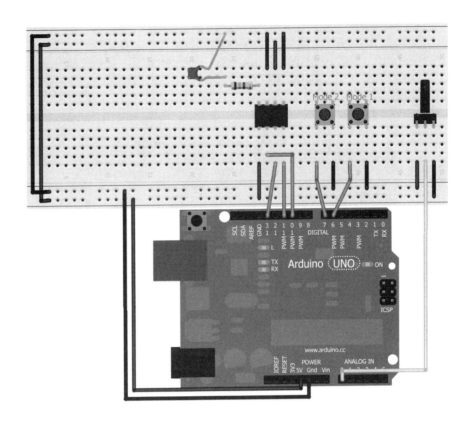

```
#define ss_pin 10
#define Channel 0
#define PPMout 4
#define mode1 6
#define mode2 7
int sensorValue = 5;

boolean bufferVref = false;
boolean port_write = true; //set false if using ss_pin other
then 10
boolean gain2x = false;

void pulseOut(byte pin, int timeMs)
{
  if (timeMs > 0){

    digitalWrite(pin, HIGH);
    delayMicroseconds(timeMs);
    digitalWrite (pin, LOW);
  }
}

void setup()
{
  Serial.begin(9600);
  pinMode(PPMout, OUTPUT);
```

```
    pinMode(mode1,INPUT_PULLUP);
    pinMode(mode2,INPUT_PULLUP);
    pinMode(ss_pin, OUTPUT);
    digitalWrite(ss_pin,HIGH);

    SPI.begin();
    SPI.setBitOrder(MSBFIRST);
    SPI.setDataMode(SPI_MODE0);
    SPI.setClockDivider(SPI_CLOCK_DIV16);
}

void loop()
{
    if (digitalRead(mode1))
    {
        if (digitalRead(mode2))
        {
            sensorValue = analogRead(0);
            sensorValue = map(sensorValue, 0, 1023, 0, 511);
            Serial.print(sensorValue);
            dispBinaryMSB(sensorValue);
            //Serial.println(sensorValue);
        }
        else
        {
            sensorValue = analogRead(0);
            sensorValue = map(sensorValue, 0, 1023, 0, 511);
            dispBinary(sensorValue);
            //Serial.print(sensorValue);
            //Serial.println(sensorValue);
        }
    }
    else if (digitalRead(mode2))
    {
        sensorValue = analogRead(0);
        sensorValue = map(sensorValue, 0, 1023, 0, 5);
        DACoutput(sensorValue*819);
        Serial.print(sensorValue);
        delay(3);

    }
    else
    {
        sensorValue = analogRead(0);
        DACoutput(sensorValue*4);
    }
}

void StartBit()
{
    pulseOut(PPMout,1000); //Start Bit
    delayMicroseconds(500);
```

```
    digitalWrite(PPMout,LOW);
}

void dispBinary(int n)
{
  StartBit();
  for (int i=8; i>=0; i--)
  {
    pulseOut(PPMout, 500*(n & 1));
    //Serial.println(500*(n & 1));
    delayMicroseconds(1000-500*(n & 1));
    n /= 2;
  }
  digitalWrite(PPMout,LOW);
  StartBit();
  delay(3);
}

void dispBinaryMSB(int n)
{
  int mosi = 256;
  StartBit();
  for (int i=8; i>=0; i--)
  {
    pulseOut(PPMout, 500*int(((n & mosi))/mosi));
    //Serial.print(500*int(((n & mosi))/mosi));
    //Serial.print(",");
    delayMicroseconds(1000-500*((n & mosi)/mosi));
    mosi /= 2;
  }
  digitalWrite(PPMout,LOW);
  StartBit();
  delay(3);
}

void DACoutput(int data) {
  // Truncate the unused bits to fit the 8/10/12 bits the DAC
  accepts
  if (bitwidth == 12)
    data &= 0xfff;
  else if (bitwidth == 10)
    data &= 0x3ff;
  else if (bitwidth == 8)
    data &= 0xff;

  if (port_write)
    PORTB &= 0xfb; // Clear PORTB pin 2 = arduino pin 10
  else
    digitalWrite(ss_pin, LOW);

  uint16_t out = (Channel << 15) | (bufferVref << 14) |
  ((!gain2x) << 13) | (1 << 12) | (data << (12 - bitwidth)));
```

```
    SPI.transfer((out & 0xff00) >> 8);
    SPI.transfer(out & 0xff);

    if (port_write)
      PORTB |= (1 << 2); // set PORTB pin 2 = arduino pin 10
    else
      digitalWrite(ss_pin, HIGH);
}
```

EXPERIMENT 18: FREQUENCY-SHIFT KEYING (FSK) MODULATION AND DEMODULATION

Breadboard diagram for FSK modulator circuit:

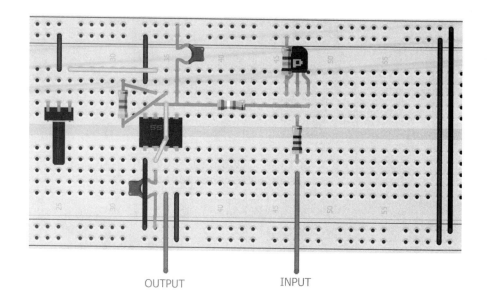

OUTPUT   INPUT

Breadboard diagram for FSK modem:

Arduino code:

```
//Pin 3 for Input Pin 13 for output

#define FSKIN 3
int duration = 0;
int center = 0;
int tolerance = 0;
void setup() {
  pinMode(FSKIN, INPUT);
  pinMode(13,OUTPUT);
  Serial.begin(115200);
  delay(100);
  //Calibration
  for (int i=0;i < 50;i++)
```

```
  {
    duration += pulseIn(FSKIN,HIGH);
  }
  center = duration / 50;
}

void loop() {
  duration = pulseIn(FSKIN, HIGH);
  if (duration > center)
  {
    digitalWrite(13,HIGH);
  }
  else
  {
    digitalWrite(13,LOW);
  }
}
```